# STATISTICAL

# GREEN'S FUNCTIONS

## V.I. Yukalov

**Queen's Papers
in Pure and Applied Mathematics
Volume 110,  1998
Kingston Ontario Canada**

# QUEEN'S PAPERS IN PURE AND APPLIED MATHEMATICS

## VOLUME 110

EDITORS:  O. BOGOYAVLENSKIJ
A. J. COLEMAN
A. V. GERAMITA
P. RIBENBOIM

## QUEEN'S UNIVERSITY, KINGSTON, ONTARIO, CANADA

1998

ISBN - 088911-824-8

Dedicated to A.J. Coleman<br>
on the occasion of his 80-th birthday

# Preface

Green's functions are widely used in various branches of mathematics and theoretical physics. Despite its rather long history, the method of Green's functions continues to develop. This development concerns first of all the application of this technique to statistical mechanics. The latter is, probably, the most complicated subject which the method of Green's functions has been applied to. Generally speaking, practically all parts of theoretical physics can be treated as subsections of statistical mechanics (this statement is explained more precisely in Appendix). It is just the complexity of statistical mechanics that perpetually forces us to develop the method of Green's functions.

There exists a volumnous literature on this method and its various applications. In no way my aim was to give a review of all that. The aim of this work is to describe the basic mathematical properties of Green's functions used in statistical mechanics, as well as the equations defining these functions and the techniques of solving these equations. No particular applications are touched in order not to be distracted from the main aim of analysing the underlying mathematical structure. Such a concentrated analysis of the fundamentals of the method may, to my mind, help to grasp its most important features and to better understand its strong and weak points. This could also give hints of how to improve and develop the approach for overcoming existing difficulties. One such a development that has not yet been published in a review form is presented here in Part 3. This is *Correlated Iteration Theory* that differs from all other known variants of perturbation theory for Green's functions by the combination of two factors: the systematic formulation of an algorithm for obtaining subsequent approximations and the consistent consideration of interparticle correlations at each step of the procedure.

The formalism of the correlated iteration theory is based on general properties of Green's functions, which are expounded in Part 2. Without studying these general properties it is impossible to proceed further. And, in turn, to introduce Green's functions for statistical systems and to study the properties of these functions, it is necessary to have those main definitions that make the foundation of statistical mechanics. These basic notions are formulated in Part 1.

Statistical Green's functions are a natural generalization of correlation functions and reduced density matrices. The relation between these objects is repeatedly traced throughout the text.

i

## Acknowledgements

Many facts about reduced density matrices as well as a lot of important things in many–body problem I have learnt from A.J. Coleman with whom I had a pleasure of enjoying numerous discussions. I greatly benefited from a lucky opportunity of having a close collaboration with A.J. Coleman, to whom I am indebted so much.

I highly appreciate permanent support and help from E.P. Yukalova.

During four years I worked at the Department of Mathematics and Statistics of Queen's University. I am very grateful to my colleagues from this Department for their friendly support and advice.


Kingston–London, May 1998.

# Contents

# 3. Correlated Iteration



# Appendix

# References

# 1 Basic Notions

This part contains those main definitions which statistical mechanics is based on and which are necessary for understanding the following parts.

## 1.1 Pure States

Consider a system of $N$ objects which we shall traditionally call particles. Put into correspondence to each particle a variable $x \in \mathbb{V}$ whose particular nature is, for the time being, of no importance. This can be a multivariable incorporating the real–space coordinates, spin, isospin and other degrees of freedom, such as colour and flavour.

For dealing with $N$ particles, we need $N$ such multivariables. Enumerating the variables (but not the particles) by the index $i = 1, 2, \ldots, N$, we have the set of $N$ variables $x_i \in \mathbb{V}_i$. For brevity, we shall use the notation

$$x^n := \{x_i|\ i = 1, 2, \ldots, n \leq N\} \tag{1.1}$$

for the set defined on the manifold

$$\mathbb{V}^n := \mathbb{V}_1 \times \mathbb{V}_2 \times \ldots \times \mathbb{V}_n \, . \tag{1.2}$$

In addition to variables (1.1) we need one more variable

$$t \in \mathbb{R}^+ := [0, \infty)$$

called *time*.

Let functions $\Psi(x^N, t)$, given on the domain $\mathbb{V}^N \times \mathbb{R}^+$, pertain to a Hilbert space $\mathcal{H}_\Psi$ with the scalar product

$$(\Psi, \Psi') := \int \Psi^*(x^N, t)\Psi'(x^N, t)dx^N, \tag{1.3}$$

in which $\Psi^*$ means complex conjugate of $\Psi$, integration goes over $\mathbb{V}^N$ and the differential measure on (1.2) is denoted by

$$dx^n := dx_1 dx_2 \ldots dx_n \, .$$

As usual, the integration implies an actual integration over continuous manifolds and summation over discrete ones.

If the elements of the Hilbert space $\mathcal{H}_\Psi$ are normalized,

$$(\Psi, \Psi) = 1, \tag{1.4}$$

then the functions $\Psi$ are called *wave functions* or *quantum states*; respectively, the space $\mathcal{H}_\Psi$ is the *space of quantum states*. The quantity

$$|\Psi(x^N, t)|^2 := \Psi^*(x^N, t)\Psi(x^N, t) \tag{1.5}$$

is interpreted as the *distribution of particle density.*

The *particles* are called *identical* or *indistinguishable* if their corresponding variables pertain to the same manifold, $\mathbb{V}_i = \mathbb{V}$, and there is no difference which variable is ascribed to which particle, so that the particle–density distribution (1.5) does not change under the permutation of any two variables:

$$|P(x_i, x_j)\Psi(x^N, t)|^2 = |\Psi(x^N, t)|^2, \tag{1.6}$$

where $P(x_i, x_j)$ is the operator permuting the variables $x_i, x_j \in \mathbb{V}$. Emphasize that it only makes sense of permuting variables but not particles.

The simplest sufficient condition for (1.6) is the *symmetry principle*

$$P(x_i, x_j)\Psi(x^N, t) = \pm\Psi(x^N, t), \tag{1.7}$$

stating that the quantum states associated with identical particles are either symmetric or antisymmetric with respect to the permutation of variables. Symmetric quantum states describe *Bose particles* or *bosons* while antisymmetric quantum states, *Fermi particles* or *fermions.*

A set $\{A\}$ of self–adjoint linear operators $A(x^N)$ is given on the space of quantum states $\mathcal{H}_\Psi$, such that the mathematical expectations

$$\langle A \rangle := (\Psi, A\Psi) \tag{1.8}$$

define observable quantities. A linear algebra of these operators forms the *algebra of observables* $\{A\}$. The set $\{\langle A \rangle\}$ of mathematical expectations (1.8) is called *observable state.*

One of the operators from the algebra of observables plays a special role. This is a self–adjoint operator $H$ whose mathematical expectation gives the energy of the system. The operator $H$, called *Hamiltonian*, defines the *evolution equation*

$$i\hbar\frac{\partial}{\partial t}\Psi(x^N, t) = H(x^N)\Psi(x^N, t) \tag{1.9}$$

for quantum states. This is the *Schrödinger equation*, $\hbar$ being the Planck constant.

Using wave functions one may construct another type of objects, called density matrices [1-3]. Introduce the *pure density matrix*

$$D_N(t) = \left[D_N(x^N, y^N, t)\right], \tag{1.10}$$

whose elements

$$D_N(x^N, y^N, t) := \Psi(x^N, t)\Psi^*(y^N, t) \tag{1.11}$$

are products of wave functions satisfying the symmetry principle (1.7). Consequently,

$$P(x_i, x_j)D_N(x^N, y^N, t) = \pm D_N(x^N, y^N, t). \tag{1.12}$$

The trace of (1.10), in compliance with (1.4), is

$$Tr \, D_n(t) := (\Psi, \Psi) = 1. \tag{1.13}$$

Also, introduce the operator matrices

$$A_N = [A_N(x^N, y^N)] \tag{1.14}$$

with elements

$$A_N(x^N, y^N) := A(x^N)\delta(x^N - y^N), \tag{1.15}$$

where

$$\delta(x^n - y^n) := \delta(x_1 - y_1)\delta(x_2 - y_2)\ldots\delta(x_n - y_n)$$

is a product of the Dirac $\delta$–functions for continuos variables and of the Kroneker $\delta$–symbols for discrete. Using (1.10) and (1.14), the mathematical expectation (1.8) is written as

$$\langle A \rangle = Tr \, A_N D_N(t). \tag{1.16}$$

The evolution equation for (1.10) follows from the Schrödinger equation (1.9),

$$i\hbar\frac{d}{dt}D_N(t) = [H_N, D_N(t)], \tag{1.17}$$

where $[\,,\,]$ means the commutator

$$[A, B] := AB - BA \,.$$

Eq. (1.17) is the *Liouville equation*. According to (1.17), the evolution equation for (1.16) is

$$i\hbar\frac{d}{dt}\langle A \rangle = i\hbar\langle\frac{\partial A}{\partial t}\rangle + \langle[A, H]\rangle. \tag{1.18}$$

The pure density matrix (1.10) and operators of observables (1.14) permit several representations. To pass from the $x$–representation given by (1.11) and (1.15) to another representation, one chooses a complete basis $\{\Psi_q(x^N)\}$ of orthonormal quantum states in $\mathcal{H}_\Psi$,

$$(\Psi_q, \Psi_p) = \delta_{qp} \,,$$

where $q$ is a multiindex labelling quantum states. With an expansion

$$\Psi(x^N, t) = \sum_q C_q(t)\Psi_q(x^N),$$

Eq. (1.4) gives

$$\sum_q |C_q(t)|^2 = 1.$$

Instead of (1.10), one may present the pure density matrix as

$$D_N(t) = [D_{qp}(t)] \tag{1.19}$$

with elements

$$D_{qp}(t) := C_q(t) C_p^*(t), \qquad (1.20)$$

which again have the form of a product, the function $C_q(t)$ playing the role of the wave function in the $q$-representation. In the mathematical expectation (1.16) one has to put now, instead of (1.14),

$$A_N = [A_{qp}] \qquad (1.21)$$

with elements

$$A_{qp} := (\Psi_q, A \Psi_p). \qquad (1.22)$$

In any representation, the pure density matrix consists of elements that are factorizable as in (1.11) or (1.20). Because of this and of normalization (1.13), the convolution property

$$\int D_N(x^N, z^N, t) D_N(z^N, y^N, t) dz^N = D_N(x^N, y^N, t) \qquad (1.23)$$

holds, which can be written in an operator form as

$$D_N^2(t) = D_N(t). \qquad (1.24)$$

The latter defines the pure density matrix as an indempotent operator.

A *pure state* is an observable state $\{\langle A \rangle\}$ consisting of mathematical expectations (1.16) defined through a pure density matrix. The *indempotency property* (1.24) is the necessary and sufficient condition of a pure state.

Reducing the number of variables in (1.11) by integrating

$$D_n(x^n, y^n, t) := \int D_N(x^n, z^{N-n}, y^n, z^{N-n}, t) dz^{N-n}, \qquad (1.25)$$

one defines the *reduced density matrices*, or submatrices,

$$D_n(t) := [D_n(x^n, y^n, t)] \qquad (n = 1, 2, \ldots, N), \qquad (1.26)$$

implying that $D_0 := 1$. It is not important which of the variables are integrated out in (1.25) since (1.11) is symmetric under the double permutation

$$P(x_i, x_j) P(y_i, y_j) D_N(x^N, y^N, t) = D_N(x^N, y^N, t).$$

The normalization for (1.26),

$$Tr\, D_n(t) := \int D_n(x^n, x^n, t) dx^n = 1, \qquad (1.27)$$

follows from (1.13). The elements (1.25) do not satisfy the convolution property (1.23), except for the trivial cases $n = 0$ or $n = N$. Thus, the reduced density matrices (1.26) do not represent indempotent operators,

$$D_n^2(t) \neq D_n(t) \qquad (n = 1, 2, \ldots, N - 1).$$

The reduced density matrices are useful for dealing with operators depending only on a part of the variables $x^N$. For example, for an operator $A(x^N)$ defining the operator matrix

$$A_n = [A_n(x^n, y^n)] \qquad (1.28)$$

with elements

$$A_n(x^n, y^n) := A(x^n)\delta(x^n - y^n), \qquad (1.29)$$

one can write the expectation (1.16) as

$$\langle A \rangle = Tr\, A_n D_n(t). \qquad (1.30)$$

In this way, to calculate (1.30) for an operator (1.28) whose elements (1.29) involve $n < N$ variables, one needs to know only the corresponding reduced density matrix (1.26). Reduced density matrices have become a very convenient tool for the many-body problem [4-14].

## 1.2  Mixed States

In reality, any considered system of $N$ particles is always a subsystem of a bigger system. Real observable systems can be never completely isolated [15-19]. Therefore, in addition to the variables $x^N$ ascribed to the subsystem of interest, there exists a set $X$ of other variables associated with the rest of a bigger system, which can be called the *background*. The whole system consisting of a subsystem and a background can be named the *universe*.

Assume that the observable state of the universe is a pure state characterized by a pure density matrix

$$D(t) = [D(x^N, X, y^N, Y, t)], \qquad (1.31)$$

whose elements

$$D(x^N, X, y^N, Y, t) := \Psi(x^N, X, t)\Psi^*(y^N, Y, t) \qquad (1.32)$$

are products of the *universe wave functions*.

The algebra of observables $\{A\}$ related to the subsystem of interest consists of operators $A(x^N)$ depending on the subsystem variables $x^N$. The mathematical expectations can be defined as in (1.16) but with the reduced density matrix $D_N(t)$, whose elements are

$$D_N(x^N, y^N, t) := \int \Psi(x^N, X, t)\Psi^*(y^N, X, t)dX . \qquad (1.33)$$

To pass to another representation, one has to choose an orthonormal basis $\{\Psi_q(x^N)\}$ and to make an expansion

$$\Psi(x^N, X, t) = \sum_q C_q(X, t)\Psi_q(x^N).$$

Then, (1.33) may be written as

$$D_N(x^N, y^N, t) = \sum_{qp} D_{qp}(t)\Psi_q(x^N)\Psi_p^*(y^N), \qquad (1.34)$$

where

$$D_{qp}(t) = \int C_q(X,t)C_p^*(X,t)dX$$

are the elements of the reduced density matrix $D_N(t)$ in the $q$–representation.

There is a sense in considering a system of $N$ particles as a subsystem of the universe solely if any of these particles has ever interacted with its surrounding background. This is just what always happens in reality. It would be an extremely unrealistic fantasy to imagine that a considered subsystem has never interacted with anything else. The fact that a subsystem cannot be completely separated from its background means that the universe wave function $\Psi(x^N, X, t)$ cannot be factorized into a product of functions each depending separately on $x^N$ and $X$. This is called the *principle of nonseparability* [13,20-22].

If the universe wave function is not factorizable in the above sense, then the elements of the reduced density matrix $D_N(t)$, given by (1.33), are not factorizable into a product of functions depending separately on $x^N$ and $y^N$, as in (1.10). Therefore (1.33) cannot define a pure density matrix.

A density matrix consisting of nonfactorizable elements will be called the *mixed density matrix*. The observable state $\{\langle A \rangle\}$, formed by mathematical expectations $\langle A \rangle$ which are defined through a mixed density matrix, is named the *mixed state*.

The necessity of characterizing any system of particles by a mixed state is not based only on the principle of nonseparability, as is described above, but can even be strengthen. Actually, the notion of an observable state itself presupposes the possibility of measuring observable quantities. But the process of measurement is impossible without interactions between the considered system and a measuring apparatus. Consequently, the latter is to be treated as a part of the background from which the system cannot be isolated.

Moreover, the concept of isolatedness of a subsystem from the rest of the universe is contradictory in its own, since to realize the isolation and to check that the subsystem is kept isolated, one again needs to use some technical and measuring devices [23,24].

Suppose that to check the isolatedness of a subsystem we use such a gentle procedure which involves only the *minimally disturbing measurement* [23,24], defined as follows: A measurement, finding the subsystem in a quantum state $\Psi_q(x^N)$, leads solely to the appearance of a phase factor $e^{i\xi_q}$,

$$\Psi_q(x^N) \rightarrow e^{i\xi_q}\Psi_q(x^N), \qquad (1.35)$$

where $\xi_q$ is a random real number, $\xi_q \in [0, 2\pi]$. The measuring process (1.35) does not change the distribution of particle density $|\Psi_q(x^N)|^2$ in the quantum state $q$, as well as the matrix elements $(\Psi_q, A\Psi_q)$. However, the total wave function

$$\Psi(x^N, \xi, t) = \sum_q C_q(t)e^{i\xi_q}\Psi_q(x^N) \qquad (1.36)$$

acquires the dependence on a set $\xi := \{\xi_q\}$ of random variables, which plays here the role of the set $X$ of background variables. To preserve the normalization

$$\int |\Psi(x^N, \xi, t)|^2 dx^N d\xi = 1, \qquad (1.37)$$

the differential measure $d\xi$ is to be defined as

$$d\xi := \prod_q \frac{d\xi_q}{2\pi}\,, \qquad \xi_q \in [0, 2\pi].$$
(1.38)

The elements of the reduced density matrix, according to (1.33), are given by

$$D_N(x^N, y^N, t) := \int \Psi(x^N, \xi, t)\Psi^*(y^N, \xi, t)d\xi\,.$$
(1.39)

The mathematical expectation $\langle A \rangle$ of an operator $A(x^N)$ is $Tr\, A_N D_N(t)$, where the elements of $A_N$ and $D_N(t)$ are defined either by (1.15) and (1.39) or by (1.22) and

$$D_{qp}(t) = |C_q(t)|^2 \delta_{qp},$$
(1.40)

respectively. The latter follows from (1.36), (1.38) and from the integral

$$\int e^{-i(\xi_q - \xi_p)}d\xi = \delta_{qp}.$$

Neither (1.39) nor (1.40) are factorizable. Hence, the subsystem is in a mixed state even for such an ideal case of the minimally disturbing measurements.

The above discussion forces us to accept the *principle of nonisolatedness* which tells that no system can be completely isolated and has to be described by a mixed density matrix. This general statement, however, does not forbid the possibility that for some particular cases a pure density matrix could be a reasonable approximation.

We shall call a *system quasi–isolated* if it is characterized by a mixed density matrix and its time evolution is governed by a Hamiltonian $H(x^N)$ depending on the system variables $x^N$, but not on the background variables $X$,

$$i\hbar\frac{\partial}{\partial t}\Psi(x^N, X, t) = H(x^N)\Psi(x^N, X, t).$$
(1.41)

The mixed density matrix of a quasi–isolated system satisfies the same Liouville equation (1.17) as a pure density matrix.

The formal solution of the Liouville equation may be written as

$$D_N(t) = U_N(t)D_N(0)U_N^+(t),$$
(1.42)

where $U_N(t)$ is the *evolution operator* which is a unitary operator obeying the Schrödinger equation

$$i\hbar\frac{d}{dt}U_N(t) = H_N U_N(t)$$
(1.43)

and having the properties

$$U_N(t)U_N(t') = U_N(t + t'),$$

$$U_N^+(t) = U_N^{-1}(t), \qquad U_N(0) = 1.$$
(1.44)

In a particular case, when $H_N$ does not depend explicitly on time, the evolution operator, according to (1.43), is

$$U_N(t) = \exp\left(-\frac{i}{\hbar}H_N(t)\right) \qquad \left(\frac{\partial H_N}{\partial t} = 0\right). \qquad (1.45)$$

With the help of the evolution operator the time evolution for the operators of observables is given as

$$A_N(t) := U_N^+(t)A_N U_N(t). \qquad (1.46)$$

The form $A_N(t)$ is named the *Heisenberg representation* for the operators, as compared to that of $A_N$ which is called the *Schrödinger representation*. From (1.46) and (1.43) the evolution equation

$$i\hbar\frac{d}{dt}A_N(t) = i\hbar\frac{\partial}{\partial t}A_N(t) + [A_N(t), H_N(t)] \qquad (1.47)$$

follows, in which

$$\frac{\partial}{\partial t}A_N(t) := U_N^+(t)\frac{\partial A_N}{\partial t}U_N(t). \qquad (1.48)$$

Eq.(1.47) is called the *Heisenberg equation*. Its important particular case is when the operator $A_N$ does not depend explicitly on time, then (1.47) and (1.48) yield

$$i\hbar\frac{d}{dt}A_N(t) = [A_N(t), H_N(t)] \qquad \left(\frac{\partial A_N}{\partial t} = 0\right). \qquad (1.49)$$

Using the properties of the evolution operator and the equality $Tr\, AB = Tr\, BA$, one can write the expectation value (1.16) in one of two forms

$$\langle A \rangle = Tr\, D_N(t)A_N = Tr\, D_N(0)A_N(t). \qquad (1.50)$$

Respectively, the evolution equation for (1.50) may be presented in two ways

$$i\hbar\frac{d}{dt}\langle A \rangle = i\hbar\langle\frac{\partial A_N}{\partial t}\rangle + Tr\, [H_N, D_N(t)]A_N =$$

$$= i\hbar\langle\frac{\partial A_N}{\partial t}\rangle + Tr\, D_N(0)[A_N(t), H_N(t)].$$

The identity of these two forms results from the properties of trace and from the relation (1.46) between the Schrödinger and Heisenberg representations.

For mixed states, the reduced density matrices can be introduced as for pure states in (1.25) and (1.26), though now there is no principal difference between $D_N(t)$ and $D_n(t)$ with $n < N$, since for a mixed state $D_N(t)$ itself is a reduced density matrix. Dealing with mixed states it is reasonable to consider all density matrices $D_n(t)$ as reduced, with the index $n$ ranging from zero, when $D_0(t) := 1$, up to infinity.

The reduced density matrices for mixed states have the symmetry property

$$P(x_i, x_j)D_n(x^n, y^n, t) = \pm D_n(x^n, y^n, t)$$

as (1.12) for pure states.

## 1.3　Field Operators

A powerful method of treating quasi–isolated systems is the field approach based on the notion of field operators. This approach is also known as the method of *second quantization*. There are several ways of describing the latter. Here we expound a variant which is, to our mind, the most general and best mathematically grounded [25].

Let $\mathcal{H}_n$ be a Hilbert space of square–integrable functions $h_n(x^n)$ with a domain $\mathbb{V}^n$ and the scalar product

$$(h_n, g_n) := \int h_n^*(x^n) g_n(x^n) dx^n,$$

where the integration goes over $\mathbb{V}^n$. Construct a Hilbert space $\mathcal{H}$ as a direct sum

$$\mathcal{H} := \oplus_{n=0}^{\infty} \mathcal{H}_n \qquad (\mathcal{H}_0 := \mathbb{C}). \tag{1.51}$$

A vector $h \in \mathcal{H}$ is presented as a column

$$h := [h_n(x^n)] = \begin{bmatrix} h_0 \\ h_1(x_1) \\ h_2(x_1, x_2) \\ \cdot \\ \cdot \\ \cdot \\ h_n(x_1, x_2, \ldots, x_n) \\ \cdot \\ \cdot \\ \cdot \end{bmatrix}. \tag{1.52}$$

The latter is a *quantum state*, or *microstate*, and $\mathcal{H}$ is a space of states or *phase space*. The scalar product in (1.51) is given by

$$(h, g) := h^+ g = \sum_{n=0}^{\infty} (h_n, g_n).$$

Define matrices of the form

$$\hat{A} := [A_{mn}(x^m, y^n)], \tag{1.53}$$

whose action on the vector (1.52) is a vector

$$\hat{A}h = \left[ \sum_{n=0}^{\infty} \int A_{mn}(x^m, y^n) h_n(y^n) dy^n \right],$$

and the multiplication of two such matrices is given by the rule

$$\hat{A}\hat{B} = \left[ \sum_{k=0}^{\infty} \int A_{mk}(x^m, z^k) B_{kn}(z^k, y^n) dz^k \right].$$

Introduce the *annihilation operator*

$$\hat{\psi}(x) = \left[\delta_{m\,n-1}\sqrt{n}\delta(x^{n-1} - y^{n-1})\delta(x - y_n)\right] \tag{1.54}$$

and the *creation operator*

$$\hat{\psi}^\dagger(x) = \left[\delta_{m\,n+1}\sqrt{n+1}\delta(x^n - y^n)\delta(x_{n+1} - x)\right], \tag{1.55}$$

having the form of the matrix (1.53) with the corresponding multiplication rules. As follows from above, the operators (1.54) and (1.55) are adjoint to each other; these are the *field operators*. Since the matrices (1.54) and (1.55) contain the Dirac $\delta$–functions, the field operators are operator distributions. The correct definition of distribution presupposes the existence of a space of square–integrable functions $f(x)$, so that

$$\hat{\psi}_f = \int f(x)\hat{\psi}(x)dx, \qquad \hat{\psi}_f^\dagger = \int f(x)\hat{\psi}^\dagger(x)dx \tag{1.56}$$

becomes operators in $\mathcal{H}$, whose domain of definition, as well as their range, contain the set of all finite vectors.

Action of the field operators on the vector (1.52) gives

$$\hat{\psi}(x)h = \left[\sqrt{n+1}h_{n+1}(x^n, x)\right],$$

$$\hat{\psi}^\dagger(x)h = \left[\sqrt{n}h_{n-1}(x^{n-1})\delta(x_n - x)\right],$$

so that $\hat{\psi}_f h$ and $\hat{\psi}_f^\dagger h$ are really finite vectors.

The explicit definition of the field operators (1.54) and (1.55) imposes no simple commutation relations on these matrices. To understand this, it is sufficient to examine their products

$$\hat{\psi}(x)\hat{\psi}(x') = \left[\delta_{m\,n-2}\sqrt{n(n-1)}\delta(x^{n-2} - y^{n-2})\delta(x - y_{n-1})\delta(x' - y_n)\right],$$

$$\hat{\psi}^\dagger(x)\hat{\psi}^\dagger(x') = \left[\delta_{m\,n+2}\sqrt{(n+1)(n+2)}\delta(x^n - y^n)\delta(x_{n+1} - x')\delta(x_{n+2} - x)\right],$$

$$\hat{\psi}(x)\hat{\psi}^\dagger(x') = \left[\delta_{mn}(n+1)\delta(x^n - y^n)\delta(x - x')\right],$$

$$\hat{\psi}^\dagger(x')\hat{\psi}(x) = \left[\delta_{mn}n\delta(x^{n-1} - y^{n-1})\delta(x_n - x')\delta(y_n - x)\right],$$

The algebraic properties of operators depend on three factors: the explicit form of an operation, the properties of the space on which the operators are defined, and the existence of some additional conditions. This is well illustrated by the example of field operators. Their explicit definition as matrices (1.54) and (1.55) does not permit to extract any simple commutation relations. Even if we narrow the domain space to a subspace enjoying some simple symmetry properties, this does not necessary lead to simple commutation rules for operators. The latter can be demonstrated by the actions

$$\hat{\psi}(x)\hat{\psi}(x')h = \left[\sqrt{(n+1)(n+2)}h_{n+2}(y^n, x, x')\right],$$

$$\hat{\psi}^{\dagger}(x)\hat{\psi}^{\dagger}(x')h = \left[\sqrt{n(n-1)}h_{n-2}(x^{n-2})\delta(x_{n-1}-x')\delta(x_n-x)\right],$$

$$\hat{\psi}(x)\hat{\psi}^{\dagger}(x')h = [(n+1)h_n(x^n)\delta(x-x')],$$

$$\hat{\psi}^{\dagger}(x')\hat{\psi}(x)h = \left[nh_n(x^{n-1},x)\delta(x_n-x')\right].$$

Suppose that we are considering a subspace of (1.51) whose vectors consist of elements having some fixed symmetry, for instance, the functions $h_n(x^n)$ are symmetric or antisymmetric. But, as is seen from above, the action of field operators does not preserve this symmetry, except for the action of the annihilation operators, $\hat{\psi}(x)\hat{\psi}(x')$. This happens because the domain of the field operators, when this domain is a subspace of (1.51), generally, does not coincide with their range. Recall that by the range of field operators we mean the range of operators (1.56).

It is clear, that to obtain simple commutation relations, one needs to invoke some additional conditions. Before doing this, let us explain why (1.54) and (1.55) are called annihilation and creation operators, respectively.

Choose as a basis of (1.51) that consisting of vectors

$$|n\rangle := [\delta_{mn}h_n(x^n)] = \begin{bmatrix} 0 \\ 0 \\ \cdot \\ \cdot \\ \cdot \\ 0 \\ h_n(x^n) \\ 0 \\ 0 \\ \cdot \\ \cdot \\ \cdot \end{bmatrix}. \tag{1.57}$$

A vector adjoint to (1.57) is a row

$$\langle n| := [0\,0\,\ldots\,h_n^*(x^n)\,0\,0\,\ldots]. \tag{1.58}$$

The notation used in (1.57) and (1.58) is called the *Dirac notation*.

The scalar product of elements from the basis $\{|n\rangle\}$ is

$$\langle m|n\rangle = \delta_{mn}(h_n, h_n).$$

So, it is a complete orthogonal basis. It is easily normalizable by putting

$$\langle n|n\rangle = (h_n, h_n) = 1.$$

Then, (1.57) is called the *n−particle state*.

The 0–particle state, that is

$$|0\rangle := [\delta_{m0}h_0] = \begin{bmatrix} h_0 \\ 0 \\ 0 \\ \cdot \\ \cdot \\ \cdot \end{bmatrix}, \tag{1.59}$$

with $|h_0| = 1$, is called the *vacuum vector* or *vacuum state*.

The action of the annihilation operator (1.54) on the vacuum vector (1.59) is

$$\hat{\psi}(x)|0\rangle = 0, \tag{1.60}$$

and its action on other vectors from the basis $\{|n\rangle\}$, with $n = 1, 2, \ldots$, gives

$$\hat{\psi}(x)|n\rangle = \sqrt{n}\left[\delta_{m\,n-1}h_n(x^{n-1}, x)\right]. \tag{1.61}$$

The action of the creation operator (1.55) on an $n$–particle vector (1.57) yields

$$\hat{\psi}^\dagger(x)|n\rangle = \sqrt{n+1}\left[\delta_{m\,n+1}h_n(x^n)\delta(x_{n+1} - x)\right]. \tag{1.62}$$

For the operators (1.56) we have from (1.60) and (1.61)

$$\hat{\psi}_f|0\rangle = 0,$$

$$\hat{\psi}_f|n\rangle = \sqrt{n}[\delta_{m\,n-1}h'_{n-1}(x^{n-1})],$$

with

$$h'_{n-1}(x^{n-1}) = \int h_n(x^{n-1}, x)f(x)dx,$$

which shows that $\hat{\psi}_f$ transforms an $n$–particle vector to an $(n-1)$–particle vector. In other words, $\hat{\psi}_f$ annihilates a particle. From (1.62) we get

$$\hat{\psi}_f^\dagger(x)|n\rangle = \sqrt{n+1}\left[\delta_{m\,n+1}h'_{n+1}(x^{n+1})\right]$$

with

$$h'_{n+1}(x^{n+1}) = h_n(x^n)f(x_{n+1}),$$

which shows that $\hat{\psi}_f^\dagger$ transforms an $n$–particle vector to an $(n+1)$–particle vector, that is, creates a particle.

In the same way one can check that the double action of the field operators, given by

$$\hat{\psi}(x)\hat{\psi}^\dagger(x')|n\rangle = (n+1)\left[\delta_{mn}h_n(x^n)\delta(x - x')\right],$$

$$\hat{\psi}^\dagger(x')\hat{\psi}(x)|n\rangle = n\left[\delta_{mn}h_n(x^{n-1}, x)\delta(x_n - x')\right],$$

means that starting from an $n$–particle vector and first creating a particle and then annihilating it, or first annihilating a particle and then creating it, one always returns to an $n$–particle vector, though not necessarily the same.

By acting $n$ times on the vacuum state (1.59), with $h_0 = 1$, by the creation operator,

$$\prod_{i=1}^{n} \hat{\psi}^{\dagger}(x_i')|0\rangle = \sqrt{n!}\left[\delta_{mn}\prod_{i=1}^{n}\delta(x_i - x_i')\right],$$

one can construct any $n$–particle vector (1.57) as

$$|n\rangle = \frac{1}{\sqrt{n!}}\int h_n(x^n)\prod_{i=1}^{n}\hat{\psi}^{\dagger}(x_i)|0\rangle dx^n \tag{1.63}$$

and any vector of space (1.51) as

$$h = \sum_{n=0}^{\infty}C_n|n\rangle,$$

where $C_n$ are complex numbers.

Till now, we have not required from the vectors of the space (1.51) any symmetry properties. But, according to the symmetry principle discussed in Section 1.1, the quantum states of identical particles are to be either symmetric or antisymmetric with respect to permutation of their variables. Symmetric quantum states correspond to bosons, antisymmetric to fermions. Thus, we have to construct the subspace of (1.51) possessing the desired symmetry properties.

The set $x^n$ of the variables $x_1, x_2, \ldots, x_n$ permits one to make $n!$ permutations. Enumerate these permutations by the index $i = 0, 1, 2, \ldots, (n-1)!$ and denote the corresponding permutation operators by $P_i(x^n)$, so that

$$P_0(x^n) = 1, \qquad P_i^2(x^n) = 1.$$

Using arbitrary functions $h_n(x^n)$, we may construct symmetric and antisymmetric functions by the receipt

$$f_n^{\pm}(x^n) = \frac{1}{\sqrt{n!}}\sum_{i=0}^{(n-1)!}(\pm 1)^i P_i(x^n)h_n(x^n). \tag{1.64}$$

By construction, from (1.64) it follows

$$P_i(x^n)f_n^{\pm}(x^n) = \pm f_n^{\pm}(x^n) \qquad (i \geq 1). \tag{1.65}$$

Denote a Hilbert space of such functions as

$$\mathcal{H}_n^{\pm} = \{f_n^{\pm}(x^n)\}.$$

Analogously to (1.51), define the Hilbert space

$$\mathcal{H}_{\pm} := \oplus_{n=0}^{\infty}\mathcal{H}_n^{\pm} \tag{1.66}$$

of vector columns

$$f_\pm = [f_n^\pm(x^n)] \in \mathcal{H}_\pm . \tag{1.67}$$

The symmetric space $\mathcal{H}_+$ represents Bose states while the antisymmetric space $\mathcal{H}_-$, Fermi states. A space having the structure (1.66), possessing a vacuum vector, is a *Fock space*.

The symmetric and antisymmetric spaces in (1.66) are subspaces of (1.51), and can be considered as projections

$$\mathcal{H}_\pm = P_\pm \mathcal{H} \tag{1.68}$$

of $\mathcal{H}$ by means of projectors $P_+$ and $P_-$.

Introduce the field operators

$$\psi_\pm(x) := P_\pm \psi(x) P_\pm,$$

$$\psi_\pm^\dagger(x) := P_\pm \psi^\dagger(x) P_\pm, \tag{1.69}$$

such that $\psi_{f+}$ and $\psi_{f+}^\dagger$, defined as in (1.56), always act inside the symmetric space $\mathcal{H}_+$, while $\psi_{f-}$ and $\psi_{f-}^\dagger$, inside the antisymmetric space $\mathcal{H}_-$. By this definition

$$\psi_+(x)\psi_-(x') = \psi_+(x)\psi_-^\dagger(x') = 0. \tag{1.70}$$

In what follows, for simplicity of notation, we shall write $\psi(x)$, representing either $\psi_+(x)$ or $\psi_-(x)$ from (1.69), and $\psi^\dagger(x)$ being either $\psi_+^\dagger(x)$ or $\psi_-^\dagger(x)$. Respectively, we will write $f$ instead of $f_\pm$ from (1.67). To distinguish a commutator from an anticommutator, we denote them as

$$[\hat{A}, \hat{B}]_\mp := \hat{A}\hat{B} \mp \hat{B}\hat{A}.$$

Using the formulas of this Section for the field operator (1.54) and (1.55), for the operators (1.69) we obtain the *commutation relations*

$$[\psi(x), \psi(x')]_\mp f = 0,$$

$$[\psi^\dagger(x), \psi^\dagger(x')]_\mp f = 0,$$

$$[\psi(x), \psi^\dagger(x')]_\mp f = \delta(x - x')f, \tag{1.71}$$

the upper sign being related to Bose– and the lower, to Fermi–cases. One also says that the commutators in (1.71) define *Bose statistics* while the anticommutator, *Fermi statistics*. Since the relations (1.71) are valid for all corresponding vectors from (1.66), a vector $f$ in (1.71) can be omitted, for example

$$[\psi(x), \psi^\dagger(x')]_\mp = \delta(x - x')\hat{1},$$

where the unity matrix is

$$\hat{1} := [\delta_{mn}\delta(x^n - y^n)].$$

Usually, for brevity, one omits the unity matrix as well.

Recall again that $\psi(x)$ and $\psi^\dagger(x)$ are operator distributions. Actual field operators are given, as in (1.56) by

$$\psi_f = \int f(x)\psi(x)dx, \qquad \psi_f^\dagger = \int f(x)\psi^\dagger(x)dx. \tag{1.72}$$

The commutation relations for (1.72) are

$$[\psi_f, \psi_g]_\mp = [\psi_f^\dagger, \psi_g^\dagger]_\mp = 0,$$

$$[\psi_f, \psi_g^\dagger]_\mp = (f, g)$$

for any square–integrable functions $f(x)$ and $g(x)$.

The time evolution of the field operators is governed by a Hamiltonian $\hat{H}$ which is a matrix of the form (1.53). The *evolution operator* $U(t)$ is defined, analogously to (1.43) and (1.44), as a unitary operator satisfying the Schrödinger equation

$$i\hbar\frac{d}{dt}U(t) = \hat{H}U(t) \tag{1.73}$$

and possessing the properties

$$U(t)U(t') = U(t + t'),$$

$$U^+(t) = U^{-1}(t) = U(-t), \tag{1.74}$$

$$U(0) = 1.$$

An annihilation field operator at time $t$ is given, as in (1.46), by the relation

$$\psi(x, t) := U^+(t)\psi(x, 0)U(t), \tag{1.75}$$

in which

$$\psi(x, 0) := \psi(x). \tag{1.76}$$

In the same way, the creation field operator $\psi^\dagger(x, t)$ is defined.

The evolution equation for $\psi(x, t)$ follows from (1.75),

$$i\hbar\frac{\partial}{\partial t}\psi(x, t) = [\psi(x, t), \hat{H}(t)] \tag{1.77}$$

where $[,\,]$ means a commutator and

$$\hat{H}(t) := U^+(t)\hat{H}U(t).$$

The equation (1.77), together with its Hermitian conjugate giving the evolution equation for $\psi^\dagger(x, t)$, are the *Heisenberg equations* for the field operators.

## 1.4   Local Observables

The operators corresponding to observable quantities are constructed with the help of the field operators. In doing this, one has to keep in mind that the manifold $\mathbb{V}$ can be unbounded. Therefore, first one has to consider submanifolds of $\mathbb{V}$ which are open bounded manifolds. Let $\Lambda \subset \mathbb{V}$ be such a submanifold. The general structure of an *operator of a local observable* is

$$\hat{A}_\Lambda := \sum_{m,n=0}^{\infty} \frac{1}{\sqrt{m!n!}} \int_\Lambda \psi^\dagger(x_m) \ldots \psi^\dagger(x_1) A(x^m, y^n) \psi(y_1) \ldots \psi(y_n) dx^m dy^n, \qquad (1.78)$$

where $A(x^m, y^n)$ can be a generalized function or an operator defined on functions with the domain $\Lambda^m \times \Lambda^n$. Integration in (1.78) for all variables goes over the same submanifold $\Lambda$.

For example, the *local number–of–particles operator* is

$$\hat{N}_\Lambda = \int_\Lambda \psi^\dagger(x)\psi(x)dx. \qquad (1.79)$$

Using the *density–of–particle operator*

$$\hat{n}(x) := \psi^\dagger(x)\psi(x), \qquad (1.80)$$

(1.79) can be written as

$$\hat{N}_\Lambda = \int_\Lambda \hat{n}(x)dx.$$

The action of (1.80) on an $n$–particle vector (1.57) gives

$$\hat{n}(x)|n\rangle = n\left[\delta_{mn} f_n(x^{n-1}, x)\delta(x_n - x)\right],$$

from where

$$\hat{N}_\Lambda |n\rangle = n|n\rangle.$$

The $n$–particle vectors are eigenvectors of $\hat{N}_\Lambda$, the numbers $n = 0, 1, 2, \ldots$, being eigenvalues. This explains why (1.79) is called the number–of–particle operator.

The time evolution of operators (1.78) follows from (1.75) giving

$$\hat{A}_\Lambda(t) = U^+(t)\hat{A}_\Lambda U(t). \qquad (1.81)$$

The evolution equation for (1.81), in compliance with (1.73), is the Heisenberg equation

$$i\hbar \frac{d}{dt}\hat{A}_\Lambda(t) = i\hbar \frac{\partial}{\partial t}\hat{A}_\Lambda(t) + [\hat{A}_\Lambda(t), \hat{H}(t)], \qquad (1.82)$$

where

$$\frac{\partial}{\partial t}\hat{A}_\Lambda(t) := U^+(t)\frac{\partial \hat{A}_\Lambda}{\partial t}U(t), \qquad (1.83)$$

which is equivalent to (1.47) and (1.48).

Introduce the *statistical operator*, or *density operator*, $\hat{\rho}(t)$, which is a positive bounded self–adjoint operator, normalized to unity,

$$Tr\,\hat{\rho}(t) = 1,\qquad(1.84)$$

and satisfying the Liouville equation

$$i\hbar\frac{d}{dt}\hat{\rho}(t) = [\hat{H}(t),\hat{\rho}(t)].\qquad(1.85)$$

The time evolution, following from (1.85), is

$$\hat{\rho}(t) = U(t)\hat{\rho}U^{+}(t);\qquad \hat{\rho} := \hat{\rho}(0).\qquad(1.86)$$

The mathematical expectations for the operators of local observables are called the *statistical averages* and are defined as

$$\langle\hat{A}_\Lambda\rangle := Tr\,\hat{\rho}(t)\hat{A}_\Lambda = Tr\,\hat{\rho}\hat{A}_\Lambda(t),\qquad(1.87)$$

which is analogous to (1.50). The trace in (1.87), as well as in (1.84), is taken over any basis from either $\mathcal{H}_+$ or $\mathcal{H}_-$ according to the statistics of particles. Since the average (1.87) depends on time, we may write

$$\langle\hat{A}_\Lambda\rangle = \langle\hat{A}_\Lambda(t)\rangle.$$

Define the *statistical density matrices* through their elements

$$\rho_{mn}(x^m,y^n,t) = Tr\,\psi(x_1)\dots\psi(x_m)\hat{\rho}(t)\psi^\dagger(y_n)\dots\psi^\dagger(y_1).\qquad(1.88)$$

Then the average (1.87) takes the form

$$\langle\hat{A}_\Lambda\rangle = \sum_{m,n=0}^{\infty}\frac{1}{\sqrt{m!n!}}\int_\Lambda A(x^m,y^n)\rho_{mn}(y^n,x^m,t)dx^m dy^n.\qquad(1.89)$$

The widespread situation is when

$$A(x^m,y^n) = \delta_{mn}A(x^n,y^n).$$

Then we need only the elements (1.88) that are diagonal with respect to the indices $m,n$, that is

$$\rho_n(x^n,y^n,t) := \rho_{nn}(x^n,y^n,t).\qquad(1.90)$$

In this case, (1.89) reads

$$\langle\hat{A}_\Lambda\rangle = \sum_{n=0}^{\infty}\frac{1}{n!}\int_\Lambda A(x^n,y^n)\rho_n(y^n,x^n,t)dx^n,dy^n.\qquad(1.91)$$

A linear space

$$\mathcal{A}_\Lambda := \{\hat{A}_\Lambda|\,\Lambda\subset\mathbb{V}\}\qquad(1.92)$$

of the self–adjoint operators of local observables is an *algebra of local observables* [26,27].

A set

$$\langle \mathcal{A}_\Lambda \rangle := \{\langle \hat{A}_\Lambda \rangle\}, \tag{1.93}$$

formed by the statistical averages (1.89), is called the *statistical state.*

If the manifold $\mathbb{V}$ is unbounded, we can define an ordered sequence $\{\Lambda_1 \subset \Lambda_2 \subset \ldots \subset \mathbb{V}\}$ of open submanifolds $\Lambda_k$, such that

$$\lim_{k \to \infty} \Lambda_k = \mathbb{V},$$

and define a set of algebras, $\{\mathcal{A}_{\Lambda_1} \subset \mathcal{A}_{\Lambda_2} \subset \ldots \mathcal{A}_{\Lambda_k} \subset \ldots\}$. The inductive limit

$$\mathcal{A}_V := \lim_{k \to \infty} \mathcal{A}_{\Lambda_k} \tag{1.94}$$

is called the *quasilocal algebra,* or *global algebra of observables.*

## 1.5 Order Indices

The statistical density matrices

$$\rho_n(t) := [\rho_n(x^n, y^n, t)] \qquad (\rho_0 := 1) \tag{1.95}$$

with elements (1.90) are matrices in the $x - y$ space. Their action on a vector

$$f_n := [f(x^n)],$$

being a column with respect to $x$, is given by the rule

$$\rho_n(t)f_n := \left[ \int \rho_n(x^n, y^n, t)f_n(y^n)dy^n \right].$$

The elements (1.90), owing to the evolution properties (1.75) and (1.86), can be written in two forms:

$$\rho_n(x^n, y^n, t) := Tr \, \psi(x_1) \ldots \psi(x_n)\hat{\rho}(t)\psi^\dagger(y_n) \ldots \psi^\dagger(y_1) =$$

$$= Tr \, \psi(x_1, t) \ldots \psi(x_n, t)\hat{\rho}\psi^\dagger(y_n, t) \ldots \psi^\dagger(y_1, t). \tag{1.96}$$

According to the commutation relations (1.71), the symmetry property

$$P(x_i, x_j)\rho_n(x^n, y^n, t) = \pm\rho_n(x^n, y^n, t)$$

holds.

The elements (1.96) cannot, in general, be factorized into products of functions depending separately on $x^n$ and $y^n$. This means that, by construction, the statistical state is a mixed state. For latter, as is discussed in Section 1.2, all density matrices with $n = 0, 1, 2, \ldots$ can be called reduced. This is especially natural when working in the field–operator representation where the number of particles is not fixed. Then one calls $\rho_n(t)$ simply an $n$–matrix.

However, we may fix the *average number of particles*

$$N := \langle \hat{N}_V \rangle = \int \rho_1(x, x, t) dx, \qquad (1.97)$$

where the integration is over $\mathbb{V}$, and

$$\rho_1(x, x, t) = \langle \hat{n}(x) \rangle = Tr\, \hat{\rho}(t) \hat{n}(x). \qquad (1.98)$$

Then, it would be reasonable to distinguish among all $n$–matrices those for which $n < N$ and to apply the name reduced only to them.

Let us fix the average number of particles (1.97) and consider a subspace of the Fock space for which

$$\langle \hat{N}_V \hat{A}_V \rangle = N \langle \hat{A}_V \rangle. \qquad (1.99)$$

Then for the 2–matrix with elements

$$\rho_2(x_1, x_2, y_1, y_2, t) = \langle \psi^\dagger(y_2) \psi^\dagger(y_1) \psi(x_1) \psi(x_2) \rangle, \qquad (1.100)$$

using the commutation relations (1.71), we can write

$$\rho_2(x_1, x_2, y_1, y_2, t) = \langle \psi^\dagger(y_2) \psi(x_2) \psi^\dagger(y_1) \psi(x_1) \rangle - \delta(x_2 - y_1) \langle \psi^\dagger(y_2) \psi(x_1) \rangle.$$

From here, taking account of (1.99), we get

$$\int \rho_2(x_1, x, x_2, x, t) dx = (N - 1) \rho_1(x_1, x_2, t). \qquad (1.101)$$

Integrating (1.101) once more, and using (1.97), we find

$$\int \rho_2(x_1, x_2, x_1, x_2, t) dx_1 dx_2 = N(N - 1).$$

Generally, the following recursion relation takes place:

$$\int \rho_n(x^{n-1}, x, y^{n-1}, x, t) dx = (N - n + 1) \rho_{n-1}(x^{n-1}, y^{n-1}, t). \qquad (1.102)$$

The normalization condition for an $n$–matrix is

$$\int \rho_n(x^n, x^n, t) dx^n = \frac{N!}{(N - n)!}. \qquad (1.103)$$

Note that the normalization (1.103) for $\rho_n(t)$ differs from the normalization (1.27) for $D_n(t)$.

Consider a system with a fixed average number of particles $N$ and occupying in the real space a volume $V$. One says that the *thermodynamic limit* is performed if

$$N \to \infty, \qquad V \to \infty, \qquad \frac{N}{V} \to const, \qquad (1.104)$$

19

which is denoted as $\lim_{N \to \infty}$ or $\lim_{V \to \infty}$.

An ensemble of particles is called a *thermodynamic system*, if the statistical state (1.93) under the thermodynamic limit (1.104) behaves so that

$$\lim_{N \to \infty} \left| \frac{1}{N} \langle \mathcal{A}_V \rangle \right| < \infty. \tag{1.105}$$

For example, the limit

$$\rho := \lim_{V \to \infty} \frac{1}{V} \langle \hat{N}_V \rangle \tag{1.106}$$

means the *average density of particles*, and the limit

$$\lim_{N \to \infty} \frac{1}{N} E_N := \lim_{V \to \infty} \frac{1}{N} \langle \hat{H}_V \rangle \tag{1.107}$$

gives the *internal energy per particle*.

To understand the behaviour of density matrices under the thermodynamic limit, define the norm

$$\|\rho_n(t)\| := \sup_{\|f_n\| \leq 1} (\rho_n(t) f_n, \rho_n(t) f_n)^{1/2}, \tag{1.108}$$

where $\rho_n(t)$ is the density matrix (1.95) and

$$\|f_n\| := (f_n, f_n)^{1/2}.$$

The statistical operator $\hat{\rho}(t)$ is, by definition, a positive self–adjoint operator. Then, the density matrix (1.95) with elements (1.96) is also positive and self–adjoint. Therefore, the norm (1.108) can also be presented as

$$\|\rho_n(t)\| = \sup_{\|f_n\| \leq 1} |(f_n, \rho_n(t) f_n)| =$$

$$= \sup_{\|f_n\| \leq 1} \int f_n^*(x^n) \rho_n(x^n, y^n, t) f_n(y^n) dx^n dy^n. \tag{1.109}$$

Define the *order indices* [28-31] for density matrices by the limit

$$\alpha_n(t) := \lim_{N \to \infty} \frac{\ln \|\rho_n(t)\|}{\ln N}, \tag{1.110}$$

where the thermodynamic limit (1.104) is assumed and $n = 0, 1, 2, \ldots$. Each density matrix $\rho_n(t)$ is positive and obeys the normalization condition (1.103), because of which, using the Stirling formula for $N \to \infty$, we have

$$0 \leq \alpha_n \leq n. \tag{1.111}$$

If the eigenvalue problem

$$\rho_n(t) \varphi_{nq}(t) = \gamma_{nq}(t) \varphi_{nq}(t)$$

is solvable for an $n$–matrix, then

$$||\rho_n(t)|| = \sup_q \gamma_{nq}(t) := \gamma_n(t),$$

and the order indices (1.110) can be expressed through the *largest eigenvalues* of the corresponding $n$–matrices:

$$\alpha_n(t) = \lim_{N\to\infty} \frac{\ln \gamma_n(t)}{\ln N}. \tag{1.112}$$

This shows the importance of the largest eigenvalues of density matrices in the thermodynamic limit.

## 1.6  Information Functionals

To define the density matrices with elements (1.96), we need to know the statistical operator $\hat{\rho}(t)$. The latter obeys the Liouville equation (1.85). This operator equation not always defines $\hat{\rho}(t)$ uniquely but some additional conditions may be required. Such conditions can be formulated by invoking the notions of information theory [32].

In what follows, implying that the operators of local observables are associated with a manifold $\mathbb{V}$, for simplicity, we shall not ascribe the corresponding index at $\hat{A}_V$ writing instead just $\hat{A}$. Also, remind that if $\hat{A}$ is an operator from the algebra $\mathcal{A}$ of local observables, and $F[\hat{A}(t)]$ is a function of $\hat{A}(t)$, then the evolution of $F[\hat{A}(t)]$ is given by the law

$$F[\hat{A}(t)] = F\left[U^+(t)\hat{A}U(t)\right] = U^+(t)F[\hat{A}]U(t),$$

where $\hat{A} := \hat{A}(0)$. This law can be considered either as a definition or it can be proved if the function $F[\hat{A}]$ is defined as a Taylor expansion in powers of $\hat{A}$. Similarly, if $F[\hat{\rho}(t)]$ is a function of the statistical operator, then, in compliance with (1.86),

$$F[\hat{\rho}(t)] = F\left[U(t)\hat{\rho}U^+(t)\right] = U(t)F[\hat{\rho}]U^+(t),$$

where $\hat{\rho} = \hat{\rho}(0)$.

Introduce the *information operator*

$$\hat{I}(t) := \ln \hat{\rho}(t). \tag{1.113}$$

The time evolution for (1.113), owing to the evolution laws discussed above, is

$$\hat{I}(t) = U(t)\hat{I}(0)U^+(t), \tag{1.114}$$

where

$$\hat{I}(0) := \ln \hat{\rho}(0) = \ln \hat{\rho}.$$

Consequently, the information operator (1.113) satisfies the Liouville equation

$$i\hbar \frac{d}{dt}\hat{I}(t) = \left[\hat{H}(t), \hat{I}(t)\right]. \tag{1.115}$$

The *Shannon information* is defined as a functional

$$I_S(t) := Tr \; \hat{\rho}(t)\hat{I}(t) = Tr \; \hat{\rho}(t)\ln\hat{\rho}(t). \tag{1.116}$$

Its relation with

$$I_S := I_S(0) = Tr \; \hat{\rho}\ln\hat{\rho} \tag{1.117}$$

easily follows from the equality

$$Tr \; \hat{\rho}(t)\hat{I}(t) = Tr \; \hat{\rho}(0)\hat{I}(0)$$

resulting from (1.86) and (1.114). Therefore, the Shannon information (1.116), actually, does not depend on time,

$$I_S(t) = I_S(0). \tag{1.118}$$

The equality (1.118) is the essence of the *Liouville theorem*.

The *information entropy* is given by

$$S := -I_S = -Tr \; \hat{\rho}\ln\hat{\rho}. \tag{1.119}$$

Since the statistical operator $\hat{\rho}$ is a positive bounded operator normalized to unity by (1.84), the information entropy (1.119) is positive and, because of (1.118), it has no time evolution.

It is important to emphasize that the Shannon information (1.116) is not an average of the information operator (1.113). An *average information* can be defined as

$$I_A(t) := Tr \; \hat{\rho}\hat{I}(t) = Tr \; \hat{\rho}\ln\hat{\rho}(t). \tag{1.120}$$

Its time evolution follows from the Liouville equation (1.115) giving

$$i\hbar\frac{d}{dt}I_A(t) = Tr \; \hat{\rho}\left[\hat{H}(t),\hat{I}(t)\right], \tag{1.121}$$

while the Shannon information, in comformity with (1.118), is constant in time. One may expect that the average information (1.120) decreases with time because of the following.

**Proposition 1.1.** (*Nonincrease of average information*).

Let the average information be given by the functional (1.120), and let an inverse operator $\hat{\rho}^{-1}(t)$ exist satisfying

$$\hat{\rho}(t)\hat{\rho}^{-1}(t) = 1,$$

then for $t \geq 0$

$$I_A(t) \leq I_A(0). \tag{1.122}$$

**Proof:** Consider the difference

$$I_A(0) - I_A(t) = Tr\hat{\rho}\left[\ln\hat{\rho} - \ln\hat{\rho}(t)\right].$$

Due to the existence of the inverse operator $\hat{\rho}^{-1}(t)$, we can write

$$\ln \hat{\rho} - \ln \hat{\rho}(t) = \ln \hat{\rho}\hat{\rho}^{-1}(t).$$

Since, by definition, $\hat{\rho}(t)$ is a positive bounded operator in a Hilbert space, its inverse is a positive operator. The product of two positive operators is a positive operator, so $\hat{\rho}\hat{\rho}^{-1}(t)$ is a positive operator. For a positive operator $\hat{X}$ in a Hilbert space the inequality

$$\ln \hat{X} \geq 1 - \hat{X}^{-1} \qquad (\hat{X} > 0)$$

holds, if the inverse operator exists. Applying this inequality to the positive operator $\hat{\rho}\hat{\rho}^{-1}(t)$, we have

$$\ln \hat{\rho}\hat{\rho}^{-1}(t) \geq 1 - \hat{\rho}^{-1}\hat{\rho}(t).$$

Consequently,

$$I_A(0) - I_A(t) \geq Tr \; \hat{\rho}\left[1 - \hat{\rho}^{-1}\hat{\rho}(t)\right].$$

As far as the statistical operator $\hat{\rho}(t)$ for any $t \geq 0$ is normalized to unity by condition (1.84), we get

$$Tr \; \hat{\rho}\left[1 - \hat{\rho}^{-1}\hat{\rho}(t)\right] = Tr \; \hat{\rho} - Tr\hat{\rho}(t) = 0.$$

Thus, we find that $Tr \; \hat{\rho}\ln \hat{\rho} \geq Tr\hat{\rho}\ln \hat{\rho}(t)$ and

$$I_A(0) - I_A(t) \geq 0,$$

which prooves (1.122).

If, instead of the entropy (1.119), we introduce the *average entropy*

$$S_A(t) := -Tr \; \hat{\rho}\ln \hat{\rho}(t), \qquad (1.123)$$

then from the above proposition, it follows

$$S_A(t) \geq S_A(0) \qquad (t \geq 0). \qquad (1.124)$$

The Shannon information (1.116) is an *unconditional information*. It may happen that some additional knowledge about the properties of a considered system is available. Then conditional information functionals can be defined.

Assume that for a set $\{\hat{A}_i(t)|\; i = 0, 1, 2, \ldots, n\}$ of operators from the algebra $\mathcal{A}$ of local observables some statistical averages

$$A_i(t) := \langle \hat{A}_i(t) \rangle = Tr \; \hat{\rho}\hat{A}_i(t) = Tr \; \hat{\rho}(t)\hat{A}_i \qquad (1.125)$$

are known. One obvious case is for the unity operator $\hat{A}_0(t) = \hat{1}$, for which we know that

$$\langle \hat{1} \rangle = Tr \; \hat{\rho}(t) = 1. \qquad (1.126)$$

Introduce the *operator of conditional information*

$$\hat{I}_{con}(t) := \ln \hat{\rho}(t) + \sum_{i=0}^{n} \lambda_i(t)\left[\hat{A}_i - A_i(t)\right] \qquad (1.127)$$

and the *conditional information*

$$I_{con}(t) := Tr\ \hat{\rho}(t)\hat{I}_{con}(t) =$$

$$= I_S + \sum_{i=0}^{n} \lambda_i(t) \left[ Tr\ \hat{\rho}(t)\hat{A}_i - A_i(t) \right], \qquad (1.128)$$

where $\lambda_i(t)$ are *Lagrange multipliers* satisfying the variational condition

$$\frac{\delta I_{con}(t)}{\delta \lambda_i(t)} = 0\ . \qquad (1.129)$$

The meaning of (1.129) is to guarantee the validity of (1.125). Note that, contrary to the Shannon information (1.116) that does not depend on time, according to (1.118), the conditional information (1.128), generally, is a function of time.

If the behaviour of some observable quantities, characterized by a set $\{\hat{A}_i(x,t)|\ i = 0,1,2,\ldots,n\}$ of operators $\hat{A}_i(x,t)$ is assumed to be known in each local vicinity of $x$ and $t$, this is described by the averages

$$A_i(x,t) := \langle \hat{A}_i(x,t) \rangle = Tr\ \hat{\rho}(t)\hat{A}_i(x). \qquad (1.130)$$

In such a case, one can define the *operator of local conditional information*

$$\hat{I}_{loc}(t) := \ln \hat{\rho}(t) + \sum_{i=0}^{n} \int \lambda_i(x,t) \left[ \hat{A}_i(x) - A_i(x,t) \right] dx \qquad (1.131)$$

and, respectively, the *local conditional information*

$$I_{loc}(t) := Tr\ \hat{\rho}(t)\hat{I}_{loc}(t), \qquad (1.132)$$

in which the variational condition

$$\frac{\delta I_{loc}(t)}{\delta \lambda_i(x,t)} = 0 \qquad (1.133)$$

defines the *Lagrange multipliers* $\lambda_i(x,t)$ guaranteeing the validity of (1.130). The local conditional information (1.132) also depends on time.

Finally, when the transition amplitudes

$$A_i(x,t,x',t') := \langle \hat{A}_i(x,t,x',t') \rangle \qquad (1.134)$$

are given as the averages of operators $\hat{A}_i(x,t,x',t')$ with the evolution property

$$\hat{A}_i(x,t,x',t') = U^+(t)\hat{A}_i(x,0,x',t')U(t),$$

then one can introduce the *operator of nonlocal conditional information*

$$\hat{I}_{non}(t) := \ln \hat{\rho}(t) + \sum_{i=0}^{n} \int \lambda_i(x,t,x',t') \times$$

$$\times \left[ \hat{A}_i(x, 0, x', t') - A_i(x, t, x', t') \right] dx\,dx'\,dt' \tag{1.135}$$

and the *nonlocal conditional information*

$$I_{non}(t) := Tr\ \hat{\rho}(t)\hat{I}_{non}(t) \tag{1.136}$$

with the Lagrange multipliers defined as in (1.129) and (1.133). Undoubtedly, (1.136) is a function of time.

When the Liouville equation (1.85) does not allow a unique definition of the statistical operator, then one can invoke the *principle of minimal information* stating that the statistical operator must have such a form that provides the minimum of information. This implies the validity of the extremum condition

$$\frac{\delta I(t)}{\delta \hat{\rho}(t)} = 0, \tag{1.137}$$

in which $I(t)$ is an information functional for either the conditional information (1.128), local conditional information (1.132), or nonlocal conditional information (1.136), depending on a case considered. Also, the inequality

$$\frac{\delta^2 I(t)}{\delta \hat{\rho}^2(t)} \geq 0 \tag{1.138}$$

is to be true, meaning that the left–hand–side of (1.138) is a positive operator. The corresponding Lagrange multipliers are given by the variational conditions as in (1.129) or (1.133). Performing the variation of information with respect to the statistical operator $\hat{\rho}(t)$, one assumes the existence of the inverse operator $\hat{\rho}^{-1}(t)$, so that

$$\frac{\delta}{\delta \hat{\rho}(t)} \ln \hat{\rho}(t) = \hat{\rho}^{-1}(t),$$

and one uses the equality

$$\frac{\delta}{\delta \hat{\rho}(t)} Tr\ \hat{\rho}(t)\hat{A}_i(t) = \hat{A}_i(t).$$

For any kind of a statistical operator, the normalization condition (1.126) is supposed to hold. So, one of the Lagrange multipliers, namely $\lambda_0(t)$, should have the same form for all statistical operators. This Lagrange multiplier may be related to the *partition function* $Z(t)$ through the equation

$$Z(t) = \exp\{1 + \lambda_0(t)\}\ . \tag{1.139}$$

In the case of the conditional information (1.128), the principle (1.137) yields

$$\hat{\rho}(t) = \frac{1}{Z(t)} \exp\left\{ -\sum_{i=1}^{n} \lambda_i(t)\hat{A}_i \right\} \tag{1.140}$$

where the partition function

$$Z(t) = Tr \exp\left\{ -\sum_{i=1}^{n} \lambda_i(t)\hat{A}_i \right\} \qquad (1.141)$$

plays the role of a normalization factor guaranteeing the validity of (1.126). The inequality (1.138) holds true since it tells that

$$\frac{\delta^2 I(t)}{\delta \hat{\rho}^2(t)} = \hat{\rho}^{-1}(t) > 0,$$

which is actually the case, as far as $\hat{\rho}(t)$ is a positive operator.

The local conditional information (1.132), in compliance with the extremum equation (1.137), leads to the *local statistical operator*

$$\hat{\rho}_{loc}(t) = \frac{1}{Z_{loc}(t)} \exp\left\{ -\sum_{i=1}^{n} \int \lambda_i(x,t)\hat{A}_i(x)dx \right\} \qquad (1.142)$$

with the *local partition function*

$$Z_{loc}(t) = Tr \exp\left\{ -\sum_{i=1}^{n} \int \lambda_i(x,t)\hat{A}_i(x)dx \right\} . \qquad (1.143)$$

In the same way, for the nonlocal conditional information (1.136) one finds the *nonlocal statistical operator*

$$\hat{\rho}_{non}(t) = \frac{1}{Z_{non}(t)} \exp\left\{ -\sum_{i=1}^{n} \int \lambda_i(x,t,x',t')\hat{A}_i(x,0,x',t')dxdx'dt' \right\} \qquad (1.144)$$

with the *nonlocal partition function*

$$Z_{non}(t) = Tr \exp\left\{ -\sum_{i=1}^{n} \int \lambda_i(x,t,x',t')\hat{A}_i(x,0,x',t')dxdx'dt' \right\} . \qquad (1.145)$$

Although the found statistical operators $\hat{\rho}(t)$, $\hat{\rho}_{loc}(t)$, and $\hat{\rho}_{non}(t)$ depend on time, however this dependence does not necessarily coincide with the time evolution prescribed by the Liouville equation (1.85). To correct this deficiency, one can do the following. Introduce the *Liouville super–operator* $\hat{\mathcal{L}}$, acting on the statistical operator $\hat{\rho}(t)$ by the rule

$$\hat{\mathcal{L}}\hat{\rho}(t) := -\frac{1}{\hbar}\left[ \hat{H}(t), \hat{\rho}(t) \right], \qquad (1.146)$$

and the operator

$$\hat{O} := \left( \frac{d}{dt} - i\hat{\mathcal{L}} \right) \hat{\rho}(t). \qquad (1.147)$$

Define the *operator of generalized conditional information*

$$\hat{I}_{gen}(t) := \hat{I}_{con}(t) + \lambda(t)\hat{O} \qquad (1.148)$$

and the *generalized conditional information*

$$I_{gen}(t) := I_{con}(t) + \lambda(t) Tr \; \hat{\rho}(t)\hat{O}. \tag{1.149}$$

Using the principle of minimal information, we obtain the *generalized statistical operator*

$$\hat{\rho}(t) = \frac{1}{Z_{gen}(t)} \exp\left\{ -\sum_{i=1}^{n} \lambda_i(t)\hat{A}_i - \lambda(t)\hat{O} \right\} \tag{1.150}$$

with the *generalized partition function*

$$Z_{gen}(t) = Tr \; \exp\left\{ -\sum_{i=1}^{n} \lambda_i(t)\hat{A}_i - \lambda(t)\hat{O} \right\}. \tag{1.151}$$

The Lagrange multiplier $\lambda(t)$ is to be defined from the variational condition

$$\frac{\delta I_{gen}(t)}{\delta \lambda(t)} = 0,$$

which yields

$$Tr \; \hat{\rho}(t)\hat{O} = Tr \; \hat{\rho}(t) \left( \frac{d}{dt} - i\hat{\mathcal{L}} \right) \hat{\rho}(t) = 0.$$

The latter equation, together with (1.146) and (1.147), gives

$$Tr \; \hat{\rho}(t) \left\{ i\hbar \frac{d}{dt}\hat{\rho}(t) - \left[ \hat{H}(t), \hat{\rho}(t) \right] \right\} = 0, \tag{1.152}$$

that is the averaged Liouville equation (1.85). In the same way, one may correct the local statistical operator (1.142) and nonlocal statistical operator (1.144).

## 1.7   Equilibrium States

The statistical state $\langle \mathcal{A} \rangle$, which is the set $\{\langle \hat{A}(t) \rangle\}$ of expectation values $\langle \hat{A}(t) \rangle$ of the operators $\hat{A}(t)$ representing observables quantities, generally, depends on time. It may happen, that this dependence is absent, so that

$$\langle \mathcal{A}(t) \rangle = \langle \mathcal{A}(0) \rangle . \tag{1.153}$$

Then (1.153) presents an *equilibrium state*, or *time–invariant state*.

Since (1.153) must hold for all operators from the algebra of observables $\mathcal{A}$, the necessary and sufficient condition for (1.153) is the time invariance of the *equilibrium statistical operator*,

$$\hat{\rho}(t) = \hat{\rho}(0) := \hat{\rho}. \tag{1.154}$$

This means that

$$\frac{d}{dt}\hat{\rho}(t) = 0, \tag{1.155}$$

and the Liouville equation (1.85) yields

$$\left[\hat{H}(t), \hat{\rho}(t)\right] = 0. \tag{1.156}$$

Eq.(1.156) is valid if and only if the statistical operator in (1.154) is a function of $\hat{H}(t)$ and of operators $\hat{C}_i(t)$, with $i = 1, 2, \ldots, n$, for which

$$\left[\hat{C}_i(t), \hat{H}(t)\right] = 0, \qquad \frac{\partial}{\partial t}\hat{C}_i(t) = 0. \tag{1.157}$$

The operators satisfying (1.157) are called the *integrals of motion*, because, according to (1.82) and (1.157),

$$\frac{d}{dt}\hat{C}_i(t) = 0, \qquad \hat{C}_i(t) = \hat{C}_i(0) := \hat{C}_i . $$

From (1.154) and (1.156) it is evident that the Hamiltonian

$$\hat{H}(t) = \hat{H}(0) := \hat{H} := \hat{C}_1 \tag{1.158}$$

is one of such integrals of motion. The expectation value

$$E := \langle \hat{H} \rangle \tag{1.159}$$

is called the *internal energy* of the system.

The knowledge that the equilibrium statistical operator (1.154) is a function of integrals of motion is not sufficient for defining its concrete form. The latter can be found with the help of the principle of minimal information as described in the previous Section. To this end, one constructs the conditional information (1.128) with the operators $\hat{A}_i$ identified with integrals of motion,

$$\hat{A}_i = \hat{C}_i \qquad (i = 1, 2, \ldots, n); \quad \hat{A}_0 = \hat{1} .$$

As a result, one comes to (1.140) and (1.141) but without time dependence,

$$\hat{\rho} = \frac{1}{Z} \exp\left(-\sum_{i=1}^{n} \lambda_i \hat{C}_i\right), \tag{1.160}$$

where

$$Z = Tr \, \exp\left(-\sum_{i=1}^{n} \lambda_i \hat{C}_i\right). \tag{1.161}$$

The statistical operator of the form (1.160) is named the *Gibbs statistical operator* and the statistical state $\langle \mathcal{A} \rangle$, composed of averages $\langle \hat{A} \rangle$ defined as $Tr \, \hat{\rho}\hat{A}$ with this statistical operator, is a *Gibbs state* [33-35].

There can be different Gibbs states depending on the given integrals of motion. For example, if the sole integral of motion is a Hamiltonian $\hat{H}$, then the conditional information (1.128) becomes

$$I_{con} = Tr \, \hat{\rho}\ln\hat{\rho} + \lambda_0 \left(Tr \, \hat{\rho} - 1\right) + \lambda_1 \left(Tr \, \hat{\rho}\hat{H} - E\right),$$

where

$$\lambda_0 := \ln Z - 1, \qquad \lambda_1 := \beta .$$

The principal of minimal information gives

$$\hat{\rho} = \frac{1}{Z} e^{-\beta \hat{H}} \tag{1.162}$$

with

$$Z = Tr \ e^{-\beta \hat{H}} . \tag{1.163}$$

The parameter $\beta$ is called the *inverse temperature*, and it is related to the *absolute temperature* measured either in energy units, $\Theta$, or in Kelvin degrees, $T$, by the equations

$$\beta \Theta = 1, \qquad \Theta = k_B T,$$

in which $k_B$ is the Boltzmann constant.

The statistical operator (1.162) is named the *Gibbs canonical operator*. One says that a physical system which it describes is represented by the *Gibbs canonical ensemble.*

A quantity $F$ satisfying the equation

$$Tr \ \exp\left(-\sum_{i=1}^{n} \lambda_i \hat{C}_i\right) = \exp\left(-\beta F\right) \tag{1.164}$$

is termed the *thermodynamic potential.* Another form of (1.164) is the definition

$$F := -\Theta \ln Tr \ \exp\left(-\sum_{i=1}^{n} \lambda_i \hat{C}_i\right) . \tag{1.165}$$

The thermodynamic potential for the Gibbs canonical ensemble, according to (1.163) and (1.165), is

$$F := -\Theta \ln Z = -\Theta \ln Tr \ e^{-\beta \hat{H}}. \tag{1.166}$$

For a system of $N$ particles inside a real–space volume $V$, the thermodynamic potential (1.166) which is a function

$$F = F(\Theta, V, N) \tag{1.167}$$

of temperature, volume, and number of particles, is called the free energy.

Writing for (1.167) the differential form

$$dF = -S_c d\Theta - p_c dV + \mu_c dN, \tag{1.168}$$

we define the *canonical entropy*

$$S_c := -\left(\frac{\partial F}{\partial \Theta}\right)_{V,N} , \tag{1.169}$$

*canonical pressure*

$$p_c := -\left(\frac{\partial F}{\partial V}\right)_{\Theta,N},$$

(1.170)

and the *canonical chemical potential*

$$\mu_c := \left(\frac{\partial F}{\partial N}\right)_{\Theta,V}.$$

(1.171)

Consider now the case of two integrals of motion, one of which is the Hamiltonian $\hat{H}$ and the second is the number–of–particles operator

$$\hat{N} = \hat{N}(t); \qquad \left[\hat{N}, \hat{H}\right] = 0.$$

(1.172)

The average number of particles is

$$N := \langle \hat{N} \rangle = Tr\, \hat{\rho}\hat{N}\ .$$

(1.173)

As the conditional information (1.128) we have

$$I_{con} = Tr\, \hat{\rho}\ln\hat{\rho} + \lambda_0\left(Tr\, \hat{\rho} - 1\right) + \lambda_1\left(Tr\, \hat{\rho}\hat{H} - E\right) + \lambda_2\left(Tr\, \hat{\rho}\hat{N} - N\right),$$

in which

$$\lambda_0 := \ln Z - 1, \qquad \lambda_1 := \beta, \qquad \lambda_2 := -\beta\mu\ .$$

The principle of minimal information yields

$$\hat{\rho} = \frac{1}{Z}e^{-\beta(\hat{H}-\mu\hat{N})}$$

(1.174)

with the partition function

$$Z = Tr\, e^{-\beta(\hat{H}-\mu\hat{N})}\ .$$

(1.175)

The parameter $\mu$ introduced above is the *chemical potential.*

The statistical operator (1.174) is called the *grand canonical operator*, and one says that a physical system described by it is represented by the *grand canonical ensemble.*

For the thermodynamic potential (1.165), we get

$$\Omega := -\Theta\ln Tr\, e^{-\beta(\hat{H}-\mu\hat{N})},$$

(1.176)

which is termed the *grand potential.* The latter is a function

$$\Omega = \Omega(\Theta, V, \mu)$$

(1.177)

of temperature, volume, and chemical potential.

The differential form

$$d\Omega = -S_g d\Theta - p_g dV - N_g d\mu$$

(1.178)

for (1.177) defines the *grand entropy*

$$S_g := - \left( \frac{\partial \Omega}{\partial \Theta} \right)_{V,\mu} , \qquad (1.179)$$

the *grand pressure*

$$p_g := - \left( \frac{\partial \Omega}{\partial V} \right)_{\Theta,\mu} , \qquad (1.180)$$

and the *grand number of particles*

$$N_g := - \left( \frac{\partial \Omega}{\partial \mu} \right)_{\Theta,V} . \qquad (1.181)$$

Similarly, one may construct other statistical operators and the corresponding thermodynamic potentials.

The quantities obtained by differentiating thermodynamic potentials are named *thermodynamic characteristics*, or *thermodynamic functions*, and their variables, *thermodynamic parameters*, or *thermodynamic variables*.

Different thermodynamic characteristics defined for different ensembles are, of course, different functions not necessary having simple relations with each other. This is because, generally, different ensembles represent different physical systems. An ensemble correctly representing a given physical system is a *representative ensemble*.

It may happen that the same system of particles is equally representable by several ensembles, which means that the analogous thermodynamic characteristics of different ensembles can be transformed to each other by a corresponding change of thermodynamic variables.

Representative ensembles equally describing the thermodynamic properties of a physical system are *thermodynamically equivalent ensembles*.

For example, if the canonical and grand canonical ensembles are thermodynamically equivalent, then the canonical entropy (1.169) and the grand entropy (1.179) are related with each other by the equations

$$S_c(\Theta, V, N_g(\Theta, V, \mu)) = S_g(\Theta, V, \mu),$$

$$S_g(\Theta, V, \mu_c(\Theta, V, N)) = S_c(\Theta, V, N),$$

the canonical pressure (1.170) and the grand pressure (1.180) satisfy the relations

$$p_c(\Theta, V, N_g(\Theta, V, \mu)) = p_g(\Theta, V, \mu),$$

$$p_g(\Theta, V, \mu_c(\Theta, V, N)) = p_c(\Theta, V, N),$$

and the relations between the canonical chemical potential (1.171) and the chemical potential $\mu$, and between the grand number of particles (1.181) and the number of particles $N$ are

$$\mu_c(\Theta, V, N_g(\Theta, V, \mu)) = \mu ,$$

$$N_g(\Theta, V, \mu_c(\Theta, V, N)) = N \ .$$

In such a case the differential forms (1.168) and (1.178) are related with each other by a Legendre transformation which gives

$$\Omega(\Theta, V, \mu) = F(\Theta, V, N_g(\Theta, V, \mu)) - \mu N_g(\Theta, V, \mu),$$

$$F(\Theta, V, N) = \Omega(\Theta, V, \mu_c(\Theta, V, N)) + \mu_c(\Theta, V, N)N.$$

Keeping this change of variables in mind, it is customary to write the latter relation as

$$\Omega = F - \mu N \ . \tag{1.182}$$

Traditionally, one also omits the indices at the thermodynamic characteristics writing

$$S = -\left(\frac{\partial F}{\partial \Theta}\right)_{V,N} = -\left(\frac{\partial \Omega}{\partial \Theta}\right)_{V,\mu},$$

$$p = -\left(\frac{\partial F}{\partial V}\right)_{\Theta,N} = -\left(\frac{\partial \Omega}{\partial V}\right)_{\Theta,\mu}, \tag{1.183}$$

$$\mu = \left(\frac{\partial F}{\partial N}\right)_{\Theta,V}, \qquad N = -\left(\frac{\partial \Omega}{\partial \mu}\right)_{\Theta,V},$$

and employing in each particular case the corresponding dependence on thermodynamic variables.

For real physical systems the thermodynamic equivalence of ensembles usually occurs only in the thermodynamic limit (1.104). Then the above relations involving the change of thermodynamic variables are to be considered as asymptotic equalities.

The thermodynamic equivalence of ensembles does not necessarily involve the equivalence of Gibbs states. This means the following. Suppose the canonical ensemble is thermodynamically equivalent to the grand canonical ensemble resulting in (1.182) and (1.183). However, the set of averages for the operators from the algebra of observables, calculated with the Gibbs canonical operator (1.162) does not necessarily coincide with the set calculated with the grand canonical operator (1.174).

Let us denote by $\langle \mathcal{A} \rangle_c$ and $\langle \mathcal{A} \rangle_g$ the Gibbs states defined through the Gibbs canonical operator (1.162) and the grand canonical operator (1.174), respectively. We shall write

$$\langle \mathcal{A} \rangle_c = \langle \mathcal{A} \rangle_g \ , \tag{1.184}$$

if all corresponding averages coincide with each other. Call $\langle \mathcal{A} \rangle_c$ the *Gibbs canonical state* and $\langle \mathcal{A} \rangle_g$ the *grand canonical state*. Then the equivalence of the statistical states (1.184) is termed the *statistical equivalence of ensembles*. The thermodynamic equivalence follows from the latter, but not reverse.

The definition of entropy given for the Gibbs canonical state in (1.169) or for the grand canonical state in (1.179) can be generalized for an arbitrary Gibbs state with a thermodynamic potential (1.165) as

$$S := -\frac{\partial F}{\partial \Theta} \ , \tag{1.185}$$

where the partial derivative with respect to temperature is taken under all other thermodynamic variables kept fixed. So defined entropy (1.185) is the *thermodynamic entropy*. Its particular examples are the canonical entropy (1.169) and the grand entropy (1.170). We may remember also that we have had the definition of the information entropy (1.119). What are conditions under which the thermodynamic and information entropies coincide?

**Proposition 1.2.** (*Criterion for entropy equivalence*).

The thermodynamic entropy (1.185) for a Gibbs state with the Gibbs statistical operator (1.160) coincides with the information entropy (1.119), that is

$$-\frac{\partial F}{\partial \Theta} = -Tr\,\hat{\rho}\ln\hat{\rho}\,, \tag{1.186}$$

if and only if the Lagrange multipliers satisfy the equality

$$\sum_{i=1}^{n}\left(\frac{\partial \lambda_i}{\partial \beta} - \frac{\lambda_i}{\beta}\right)\langle\hat{C}_i\rangle = 0\,. \tag{1.187}$$

**Proof:** From the definition of the thermodynamic entropy (1.185), with the thermodynamic potential (1.165), we have

$$S = \beta\left(\sum_{i=1}^{n}\frac{\partial \lambda_i}{\partial \beta}\langle\hat{C}_i\rangle - F\right)\,. \tag{1.188}$$

Now, let us consider the information entropy (1.119). For the statistical operator (1.160), we get

$$\ln\hat{\rho} = -\sum_{i=1}^{n}\lambda_i\hat{C}_i - \ln Z\,. \tag{1.189}$$

With the thermodynamic potential $F$ defined in (1.166), this results in

$$-Tr\,\hat{\rho}\ln\hat{\rho} = \beta\left(\sum_{i=1}^{n}\frac{\lambda_i}{\beta}\langle\hat{C}_i\rangle - F\right)\,. \tag{1.190}$$

The comparison of expressions (1.188) and (1.190) shows that they coincide if and only if the condition (1.187) holds, which concludes the proof.

**Corollary.** The canonical entropy (1.169) and the grand entropy (1.179) coincide with the information entropy (1.119) expressed in the corresponding variables.

In the case of the Gibbs canonical state,

$$\lambda_1 = \beta, \qquad \frac{\partial \lambda_1}{\partial \beta} = 1,$$

and the criterion (1.187) is valid. The thermodynamic entropy has the form

$$S = \beta\,(E - F)$$

coinciding with the information entropy.

For the grand canonical state the Lagrange multipliers are

$$\lambda_1 = \beta, \qquad \lambda_2 = -\beta\mu$$

and the criterion (1.187) holds since

$$\frac{\partial \lambda_1}{\partial \beta} = 1, \qquad \frac{\partial \lambda_2}{\partial \beta} = -\mu .$$

The thermodynamic entropy (1.188) equals the information entropy (1.119) being

$$S = \beta \left( E - \mu N - \Omega \right).$$

Note that in these two cases we have the equality

$$\frac{\partial \lambda_i}{\partial \beta} = \frac{\lambda_i}{\beta} , \tag{1.191}$$

which is a sufficient condition for (1.187)

Similarly to the considered cases, we may define other statistical states described by the Gibbs statistical operator (1.160) with other integrals of motion. For example, if the real–space volume of the system is not fixed, we can introduce the *volume operator* $\hat{V}$, which is a projector from the real space to a chosen volume, such that

$$\langle \hat{V} \rangle = V . \tag{1.192}$$

Then, constructing the corresponding conditional information and minimizing it, one comes to the *Gibbs isobaric state* with the *Gibbs isobaric operator*

$$\hat{\rho} = \frac{1}{Z} e^{-\beta(\hat{H} - p\hat{V})} , \tag{1.193}$$

in which

$$Z = Tr \; e^{-\beta(\hat{H} - p\hat{V})} .$$

The thermodynamic potential related to the Gibbs isobaric state,

$$G := -\Theta \ln Tr \; e^{-\beta(\hat{H} - p\hat{V})} , \tag{1.194}$$

is called the *Gibbs potential.* It is a function

$$G = G(\Theta, p, N) \tag{1.195}$$

of temperature, pressure and number of particles, with the differential form

$$dG = -Sd\Theta + Vdp + \mu dN . \tag{1.196}$$

The later defines the *isobaric entropy*

$$S := -\left(\frac{\partial G}{\partial \Theta}\right)_{p,N} , \qquad (1.197)$$

the *average volume*

$$V := \left(\frac{\partial G}{\partial p}\right)_{\Theta,N} , \qquad (1.198)$$

and the *isobaric chemical potential*

$$\mu := \left(\frac{\partial G}{\partial N}\right)_{\Theta,p} . \qquad (1.199)$$

In formulas (1.196) to (1.199), following tradition, the index specifying the affiliation of the thermodynamic characteristics to an ensemble, in this case to the *isobaric ensemble*, for brevity, is omitted. The relation of the Gibbs potential (1.194) with the free energy and the grand potential is

$$G = F + pV = \Omega + \mu N + pV . \qquad (1.200)$$

The general form of the thermodynamic potential (1.165) has remarkable convexity properties with respect to Lagrange multipliers. The first derivative of (1.165) is

$$\frac{\partial F}{\partial \lambda_i} = \Theta \langle \hat{C}_i \rangle .$$

Differentiating this once more, we find

$$\frac{\partial^2 F}{\partial \lambda_i^2} = -\Theta \left( \langle \hat{C}_i^2 \rangle - \langle \hat{C}_i \rangle^2 \right) .$$

For any self–adjoint operator $\hat{A}$, the quantity

$$\langle \hat{A}^2 \rangle - \langle \hat{A} \rangle^2 = \langle \left( \hat{A} - \langle \hat{A} \rangle \right)^2 \rangle$$

is nonnegative. If integrals of motion pertain to the algebra of observables, they are self–adjoint. In such a case,

$$\frac{\partial^2 F}{\partial \lambda_i^2} \le 0 \qquad (i \ge 1), \qquad (1.201)$$

that is, the thermodynamic potential is a concave function with respect to each of its Lagrange multipliers. Inequalities (1.201) are named the *conditions of thermodynamic stability*.

Defining the *specific heat*

$$C_V := \frac{1}{N}\left(\frac{\partial E}{\partial \Theta}\right)_{V,N} = \frac{\Theta}{N}\left(\frac{\partial S}{\partial \Theta}\right)_{V,N} = -\frac{\Theta}{N}\left(\frac{\partial^2 F}{\partial \Theta^2}\right)_{V,N} \qquad (1.202)$$

and the *isothermal compressibility*

$$K_T := -\frac{1}{V}\left(\frac{\partial V}{\partial p}\right)_{\Theta,N} = \frac{1}{\rho}\left(\frac{\partial \rho}{\partial p}\right)_{\Theta,N} = -\frac{1}{V}\left(\frac{\partial^2 G}{\partial p^2}\right)_{\Theta,N} \; , \qquad (1.203)$$

we see that they are expressed through the second derivatives of thermodynamic potentials, whence are to be sign–definite. Really, for the specific heat (1.202) we get

$$C_V = \frac{\beta^2}{N}\left(\langle \hat{H}^2 \rangle - \langle \hat{H} \rangle^2\right) . \qquad (1.204)$$

The isothermal compressibility (1.203) can also be found from the expressions

$$K_T^{-1} := -V\left(\frac{\partial p}{\partial V}\right)_{\Theta,N} = \rho\left(\frac{\partial p}{\partial \rho}\right)_{\Theta,N} = V\left(\frac{\partial^2 F}{\partial V^2}\right)_{\Theta,N} .$$

Finally, one obtains

$$K_T = \frac{\beta V}{N^2}\left(\langle \hat{N}^2 \rangle - \langle \hat{N} \rangle^2\right) . \qquad (1.205)$$

The quantities (1.204) and (1.205) are positively defined,

$$C_V \geq 0, \qquad K_T \geq 0 . \qquad (1.206)$$

These are the most often used stability conditions. When at least one of the inequalities in (1.206) becomes an equality, this defines the *stability boundary*.

## 1.8  Quasiequilibrium States

A *quasiequilibrium state*, or *locally equilibrium state*, is a statistical state $\langle \mathcal{A} \rangle$ formed by the operator averages with a local statistical operator [36,37] (see also [19]).

To concretize the construction of the latter, introduce the *Hamiltonian density* $\hat{H}(x,t)$ such that

$$\hat{H}(t) = \int \hat{H}(x,t)dx, \qquad (1.207)$$

and recall that the number–of–particles operator can be written as

$$\hat{N}(t) = \int \hat{n}(x,t)dx , \qquad (1.208)$$

where the density–of–particle operator

$$\hat{n}(x,t) = \psi^\dagger(x,t)\psi(x,t) \qquad (1.209)$$

is defined analogously to (1.80).

As the local averages (1.130), choose the *energy density*

$$E(x,t) := \langle \hat{H}(x,t) \rangle \qquad (1.210)$$

and the *particle density*

$$\rho(x,t) := \langle \hat{n}(x,t) \rangle \ . \tag{1.211}$$

The integration of (1.210),

$$E(t) = \int E(x,t)dx \ ,$$

gives the internal energy (1.159). The integral of (1.211) yields the average number of particles

$$N(t) = \int \rho(x,t)dx \ .$$

Note that, according to (1.98), the particle density (1.211) is a diagonal element of the reduced density matrix

$$\rho_1(x,x,t) = \rho(x,t) \ .$$

With the given local averages (1.210) and (1.211), the local conditional information (1.132) becomes

$$I_{loc}(t) = Tr \ \hat{\rho}(t) \ln \hat{\rho}(t) + \lambda_0(t)\left[Tr \ \hat{\rho}(t) - 1\right] +$$

$$+ \int \lambda_1(x,t)\left[Tr \ \hat{\rho}(t)\hat{H}(x,t) - E(x,t)\right]dx +$$

$$+ \int \lambda_2(x,t)\left[Tr \ \hat{\rho}(t)\hat{n}(x,t) - \rho(x,t)\right]dx, \tag{1.212}$$

where

$$\lambda_0(t) := \ln Z_{loc}(t) - 1, \qquad \lambda_1(x,t) := \beta(x,t), \qquad \lambda_2(x,t) := -\beta(x,t)\mu(x,t) \ .$$

The function $\beta(x,t)$ is the *local inverse temperature* connected with the *local absolute temperature* $\Theta(x,t)$ by the relation

$$\beta(x,t)\Theta(x,t) = 1 \ ,$$

and $\mu(x,t)$ is the *local chemical potential.*

The principle of minimal information for (1.212) yields the local statistical operator

$$\hat{\rho}_{loc}(t) = \frac{1}{Z_{loc}(t)} \exp\left\{ -\int \beta(x,t)\left[\hat{H}(x,t) - \mu(x,t)\hat{n}(x,t)\right]dx \right\}, \tag{1.213}$$

with the local partition function

$$Z_{loc}(t) = Tr \exp\left\{ -\int \beta(x,t)\left[\hat{H}(x,t) - \mu(x,t)\hat{n}(x,t)\right]dx \right\} \ . \tag{1.214}$$

Introduce the *quasiequilibrium potential* $Q(t)$ requiring that

$$Z_{loc}(t) = e^{-Q(t)} \ ,$$

which gives the definition

$$Q(t) := -\ln Z_{loc}(t) =$$

$$= -\ln Tr \, \exp\left\{-\int \beta(x,t)\left[\hat{H}(x,t) - \mu(x,t)\hat{n}(x,t)\right]dx\right\} . \tag{1.215}$$

One may also define the *local grand potential* $\Omega(x,t)$ by the equation

$$Z_{loc}(t) := \exp\left\{-\int \beta(x,t)\Omega(x,t)dx\right\} . \tag{1.216}$$

From (1.215) and (1.216) it follows that

$$Q(t) = \int \beta(x,t)\Omega(x,t)dx . \tag{1.217}$$

We need to find how the local grand potential changes when varying the local absolute temperature $\Theta(x,t)$ and the local chemical potential $\mu(x,t)$. To this end, we, first, obtain from (1.215) the variational derivatives

$$\frac{\delta Q(t)}{\delta\beta(x,t)} = E(x,t) - \mu(x,t)\rho(x,t), \qquad \frac{\delta Q(t)}{\delta\mu(x,t)} = -\beta(x,t)\rho(x,t).$$

Then, we find the same derivatives from (1.217),

$$\frac{\delta Q(t)}{\delta\beta(x,t)} = \Omega(x,t) + \beta(x,t)\frac{\delta\Omega(x,t)}{\delta\beta(x,t)} , \qquad \frac{\delta Q(t)}{\delta\mu(x,t)} = \beta(x,t)\frac{\delta\Omega(x,t)}{\delta\mu(x,t)} .$$

Comparing the derivatives with respect to $\beta(x,t)$, we get

$$\frac{\delta\Omega(x,t)}{\delta\beta(x,t)} = \Theta(x,t)\left[E(x,t) - \mu(x,t)\rho(x,t) - \Omega(x,t)\right] .$$

Using the relation

$$\frac{\delta\Omega(x,t)}{\delta\beta(x,t)} = -\Theta^2(x,t)\frac{\delta\Omega(x,t)}{\delta\Theta(x,t)} ,$$

we have

$$\frac{\delta\Omega(x,t)}{\delta\Theta(x,t)} = \beta(x,t)\left[\Omega(x,t) + \mu(x,t)\rho(x,t) - E(x,t)\right] .$$

Analogously to the equilibrium case (1.179), we may define the *local thermodynamic entropy*

$$S(x,t) := -\frac{\delta\Omega(x,t)}{\delta\Theta(x,t)} . \tag{1.218}$$

Substituting to the right–hand side of (1.218) the variational derivative found above, we obtain

$$S(x,t) = \beta(x,t)\left[E(x,t) - \mu(x,t)\rho(x,t) - \Omega(x,t)\right] . \tag{1.219}$$

Comparing the derivatives of $Q(t)$ with respect to $\mu(x,t)$, we find

$$\rho(x,t) = -\frac{\delta\Omega(x,t)}{\delta\mu(x,t)} . \tag{1.220}$$

In compliance with (1.218) and (1.220), the variational equation

$$\delta\Omega(x,t) = -S(x,t)\delta\Theta(x,t) - \rho(x,t)\delta\mu(x,t) \tag{1.221}$$

holds, named the *Gibbs–Duhem relation*.

The *local pressure* is defined as

$$p(x,t) := -\Omega(x,t) \ . \tag{1.222}$$

The variation of (1.219) gives two other useful relations

$$\beta(x,t) = \frac{\delta S(x,t)}{\delta E(x,t)}, \qquad \mu(x,t) = -\Theta(x,t)\frac{\delta S(x,t)}{\delta\rho(x,t)} \ ,$$

which show how the local inverse temperature and the local chemical potential are connected with the variation of the local thermodynamic entropy.

In addition to the latter, we can introduce the *quasiequilibrium information entropy*

$$S(t) := -Tr\ \hat\rho_{loc}(t)\ln\hat\rho_{loc}(t) \ . \tag{1.223}$$

With the local statistical operator (1.213), this gives

$$S(t) = \int \beta(x,t)\left[E(x,t) - \mu(x,t)\rho(x,t)\right]dx - Q(t) \ ,$$

where $Q(t)$ is defined in (1.215). Taking into account (1.217) and (1.219), we obtain the relation

$$S(t) = \int S(x,t)dx \tag{1.224}$$

between the quasiequilibrium information entropy (1.223) and the local thermodynamic entropy (1.218).

The local statistical operator (1.213), generally, does not satisfy the Liouville equation (1.85), thus, its time evolution cannot be described by a simple unitary transformation, like (1.86). Therefore, the quasiequilibrium information entropy (1.223), contrary to the information entropy (1.119), depends on time. This permits one to use the *quasiequilibrium ensemble*, with the local statistical operator (1.213), as a representation of *open*, or *non–isolated systems* whose time evolution cannot be described by the Liouville equation.

When applying the local statistical operator to quasi–isolated systems whose time evolution obeys the Liouville equation, one can correct the time behaviour of this operator by introducing one more Lagrange multiplier, as is described in Section 1.6. One may also employ the local statistical operator as a starting approximation in constructing perturbation theory for the Liouville equation. Then, the higher–order approximations of perturbation theory will automatically correct the starting approximation adjusting its time behaviour to the Liouville law. Finally, the local statistical operator $\hat\rho_{loc}(0)$ can be used as an initial condition for the Liouville equation.

## 1.9 Evolution Equations

The statistical state $\langle \mathcal{A}(t) \rangle$ which is not time–invariant is a *nonequilibrium state*. The statistical operator for which (1.154) is not valid is a *nonequilibrium statistical operator*. The time evolution of the latter is given by the Liouville equation (1.85).

However, what one finally needs is not a statistical operator itself, but the averages of operators associated with observable quantities. For an arbitrary operator $\hat{A}$, the statistical averages, as defined in (1.87), satisfy the property

$$Tr \, \hat{\rho}(t)\hat{A}(0) = Tr \, \hat{\rho}(0)\hat{A}(t) \,, \tag{1.225}$$

where $\hat{A}(t)$ is the Heisenberg representation of an operator $\hat{A}$ with the time evolution (1.81). Therefore, the equation of motion for the average (1.225) can be derived by using either the Liouville equation for the statistical operator $\hat{\rho}(t)$ or the Heisenberg equation for the operator $\hat{A}(t)$.

Keeping in mind the general case, when the number of particles in the system is not fixed, one defines the *generalized Hamiltonian*

$$H(t) := \hat{H}(t) - \mu \hat{N}(t) \,, \tag{1.226}$$

which can also be called the *grand Hamiltonian*, since this form is preferable for dealing with the grand ensemble. The time evolution is assumed to be governed by (1.226).

In order to simplify formulas, it is customary to pass to the system of units in which the repeatedly appearing Planck constant is set to unity, $\hbar = 1$.

In this notation, the Liouville equation for the statistical operator becomes

$$i\frac{d}{dt}\hat{\rho}(t) = [H(t), \hat{\rho}(t)] \,, \tag{1.227}$$

and the Heisenberg equation for an operator $\hat{A}(t)$ is

$$i\frac{d}{dt}\hat{A}(t) = i\frac{\partial}{\partial t}\hat{A}(t) + \left[\hat{A}(t), H(t)\right] \,, \tag{1.228}$$

where the partial time derivative is defined in (1.83). The evolution equations governed by a Hamiltonian define the *Hamiltonian dynamics*.

The time evolution of a statistical average $\langle \hat{A}(t) \rangle$, given by one of the forms in (1.225), can be presented by two equations. Using the left–hand side of (1.225) and (1.227), one has

$$i\frac{d}{dt}\langle \hat{A}(t) \rangle = Tr \, [H(t), \hat{\rho}(t)] \, \hat{A}(0) \,. \tag{1.229}$$

From the right–hand side of (1.225), together with (1.228), one obtains

$$i\frac{d}{dt}\langle \hat{A}(t) \rangle = \langle i\frac{\partial}{\partial t}\hat{A}(t) \rangle + \langle \left[\hat{A}(t), H(t)\right] \rangle \,. \tag{1.230}$$

These two equations are completely equivalent, and the choice of one of them is a matter of taste and convenience [38-42]. According to our own experience, the evolution equation (1.230) is usually easier to treat.

To give examples of evolution equations, we need to concretize the grand Hamiltonian (1.226). For a nonrelativistic system of particles with a two–body *interaction potential* $V(x, x')$, the generally accepted Hamiltonian is

$$H(t) := \int \psi^\dagger(x, t) \left[ K(\vec{\nabla}) + U(x, t) - \mu \right] \psi(x, t) dx +$$

$$+ \frac{1}{2} \int \psi^\dagger(x, t) \psi^\dagger(x', t) V(x, x') \psi(x', t) \psi(x, t) dx dx' , \tag{1.231}$$

where $K(\vec{\nabla})$ is a *kinetic–energy operator* with $\vec{\nabla} := \partial / \partial \, \vec{r}$, and $\vec{r} := \{ r_\alpha |\ \alpha = 1, 2, 3 \}$ being the real–space vector; the multivariable $x = \{ \vec{r}, \ldots, \}$ is assumed to include $\vec{r}$; $U(x, t)$ is an *external potential*; and $\mu$, chemical potential. The field operators satisfy the commutation relations

$$\left[ \psi(x, t) \psi^\dagger(x', t) \right]_{\mp} = \delta(x - x'), \qquad [\psi(x, t) \psi(x', t)]_{\mp} = 0, \tag{1.232}$$

which follows from (1.71) and the Heisenberg evolution (1.75), the upper sign corresponding to Bose–, and lower, to Fermi statistics. Emphasize that (1.232) holds only for the field operators at coinciding times.

From the Heisenberg equation for an annihilation operator

$$i \frac{\partial}{\partial t} \psi(x, t) = [\psi(x, t), H(t)] , \tag{1.233}$$

employing (1.231) and (1.232), one gets

$$i \frac{\partial}{\partial t} \psi(x, t) = \left[ K(\vec{\nabla}) + U(x, t) - \mu \right] \psi(x, t) +$$

$$+ \int \Phi(x, x') \psi^\dagger(x', t) \psi(x', t) \psi(x, t) dx' , \tag{1.234}$$

where there appears the *symmetrized interaction potential*

$$\Phi(x, x') := \frac{1}{2} \left[ V(x, x') + V(x', x) \right] , \tag{1.235}$$

having the property

$$\Phi(x, x') = \Phi(x', x) .$$

Note that the same equation (1.234) may be obtained from (1.231) by the variational procedure

$$i \frac{\partial}{\partial t} \psi(x, t) = \frac{\delta H}{\delta \psi^\dagger(x, t)} . \tag{1.236}$$

The time evolution of the creation operator $\psi^\dagger(x,t)$ is given by the Hermitian conjugate of (1.233)–(1.236).

To write the equation of motion for the density–of–particles operator (1.209), take the kinetic–energy operator as

$$K(\vec{\nabla}) := -\frac{\nabla^2}{2m}\,, \tag{1.237}$$

where $m$ is a *particle mass*. Introduce the *density–of–current operator*

$$\hat{\vec{j}}(x,t) := -\frac{i}{2m}\left\{\psi^\dagger(x,t)\,\vec{\nabla}\,\psi(x,t) - \left[\vec{\nabla}\,\psi^\dagger(x,t)\right]\psi(x,t)\right\}\,. \tag{1.238}$$

The evolution equation for the density–of–particles operator can be reduced to the form

$$\frac{\partial}{\partial t}\hat{n}(x,t) + \operatorname{div}\hat{\vec{j}}(x,t) = 0\,, \tag{1.239}$$

called the *continuity equation*. Here $\operatorname{div}\vec{j}\equiv\vec{\nabla}\vec{j}$.

Using again (1.234), together with its Hermitian conjugate, we may derive the equations for the components $\hat{j}_\alpha(x,t)$ of the density–of–current operator $\hat{\vec{j}}(x,t) = \{\hat{j}_\alpha(x,t)|\ \alpha = 1,2,3\}$ given in (1.238). For simplicity, consider the case of no external potentials, $U(x,t) = 0$. Then we find

$$\frac{\partial}{\partial t}\hat{j}_\alpha(x,t) + \frac{1}{2m^2}\sum_\beta \frac{\partial}{\partial r_\beta}\left\{\left[\frac{\partial}{\partial r_\alpha}\psi^\dagger(x,t)\right]\frac{\partial}{\partial r_\beta}\psi(x,t)+\right.$$

$$\left. + \left[\frac{\partial}{\partial r_\beta}\psi^\dagger(x,t)\right]\frac{\partial}{\partial r_\alpha}\psi(x,t) - \frac{1}{2}\frac{\partial^2\hat{n}(x,t)}{\partial r_\alpha\partial r_\beta}\right\} =$$

$$= -\frac{1}{m}\int\left[\frac{\partial}{\partial r_\alpha}\Phi(x,x')\right]\psi^\dagger(x,t)\psi^\dagger(x',t)\psi(x',t)\psi(x,t)dx'\,. \tag{1.240}$$

The *density–of–energy operator* is defined as

$$\hat{H}(x,t) := \frac{1}{2m}\,\vec{\nabla}\,\psi^\dagger(x,t)\cdot\vec{\nabla}\,\psi(x,t)+$$

$$+\frac{1}{2}\int V(x,x')\psi^\dagger(x,t)\psi^\dagger(x',t)\psi(x',t)\psi(x,t)dx'\,. \tag{1.241}$$

This is a particular case of the Hamiltonian density (1.207). The evolution equation for (1.241) is

$$\frac{\partial}{\partial t}\hat{H}(x,t) + \operatorname{div}\left\{\frac{i}{(2m)^2}\left[\nabla^2\psi^\dagger(x,t)\cdot\vec{\nabla}\,\psi(x,t) - \vec{\nabla}\,\psi^\dagger(x,t)\cdot\nabla^2\psi(x,t)\right]+\right.$$

$$\left. +\frac{1}{2}\int\Phi(x,x')\psi^\dagger(x,t)\hat{\vec{j}}(x,t)\psi(x',t)dx'\right\} =$$

$$= -\frac{1}{2}\int \vec{\nabla}\,\Phi(x,x')\cdot\left[\psi^\dagger(x',t)\hat{\vec{j}}(x,t)\psi(x',t)+\psi^\dagger(x,t)\hat{\vec{j}}(x',t)\psi(x,t)\right]dx' \ . \qquad (1.242)$$

The evolution equations (1.239), (1.240), and (1.242) are called the *local conservation laws*. Accomplishing statistical averaging of these laws, one derives the equations of motion for observables quantities as in (1.230).

Another way of describing statistical states is based on the fact that these states are completely defined by the reduced density matrices with elements (1.96). For example, the average number of particles (1.97) involves only 1–matrix. The internal energy (1.159), with the Hamiltonian $\hat{H}(t)$ given by (1.226) and (1.231), that is

$$E(t) = \langle \hat{H}(t)\rangle = \langle H(t) + \mu\hat{N}\rangle \ , \qquad (1.243)$$

can be expressed through 1– and 2–matrices,

$$E(t) = \int K(x_1,x_2,t)\rho_1(x_2,x_1,t)dx_1dx_2+$$

$$+\frac{1}{2}\int V(x_1,x_2)\rho_2(x_2,x_1,x_2,x_1,t)dx_1dx_2 \ , \qquad (1.244)$$

where the notation for the *one–particle kernel*

$$K(x_1,x_2,t) := \delta(x_1-x_2)K_1(x_2,t)$$

and the *one–particle reduced Hamiltonian*

$$K_1(x,t) := K(\vec{\nabla}) + U(x,t) \qquad (1.245)$$

are introduced. When the number of particles, $N$, is fixed, then, using relation (1.101), one can cast (1.244) into the form

$$E(t) = \int \delta(x_1-y_1)\delta(x_2-y_2)\frac{K_2(x_1,x_2,t)}{2(N-1)}\rho_2(x_1,x_2,y_1,y_2,t)dx_1dx_2dy_1dy_2 \ , \qquad (1.246)$$

with the *two–particle reduced Hamiltonian*

$$K_2(x_1,x_2,t) := K_1(x_1,t) + K_1(x_2,t) + (N-1)V(x_1,x_2) \ . \qquad (1.247)$$

To find the internal energy (1.246), one needs to know only 2–matrix.

Generally, all averages corresponding to observable quantities can be expressed through several, usually a few, low–order density matrices. Therefore, one has to find out the evolution equations for the latter. These equations can be derived by using (1.234). Thus, for 1–matrix we have

$$i\frac{\partial}{\partial t}\rho_1(x_1,x_2,t) = [K_1(x_1,t) - K_1(x_2,t)]\rho_1(x_1,x_2,t)+$$

$$+\int [\Phi(x_1,x_3) - \Phi(x_2,x_3)]\rho_2(x_3,x_1,x_3,x_2,t)dx_3 \ . \qquad (1.248)$$

For 2–matrix, we obtain

$$i\frac{\partial}{\partial t}\rho_2(x_1, x_2, y_1, y_2, t) = [H(x_1, x_2, t) - H(y_1, y_2, t)]\,\rho_2(x_1, x_2, y_1, y_2, t)+$$

$$+ \int [\Phi(x_1, x_2, x_3) - \Phi(y_1, y_2, x_3)]\,\rho_3(x_3, x_1, x_2, x_3, y_1, y_2, t)dx_3 \ , \qquad (1.249)$$

where the *two–particle Hamiltonian*

$$H(x_1, x_2, t) := K_1(x_1, t) + K_1(x_2, t) + \Phi(x_1, x_2) \ , \qquad (1.250)$$

and the *three–particle interaction*

$$\Phi(x_1, x_2, x_3) := \Phi(x_1, x_3) + \Phi(x_2, x_3)$$

are introduced.

If the kinetic–energy operator has the form of (1.237), then, using the property

$$\nabla_1^2 - \nabla_2^2 = \left(\vec{\nabla}_1 + \vec{\nabla}_2\right)\left(\vec{\nabla}_1 - \vec{\nabla}_2\right) ,$$

the 1–matrix equation (1.248) can be written as

$$i\frac{\partial}{\partial t}\rho_1(x_1, x_2, t) = -i\left(\vec{\nabla}_1 + \vec{\nabla}_2\right)\vec{j}\,(x_1, x_2, t) + [U(x_1, t) - U(x_2, t)]\,\rho_1(x_1, x_2, t)+$$

$$+ \int [\Phi(x_1, x_3) - \Phi(x_2, x_3)]\,\rho_2(x_3, x_1, x_3, x_2, t)dx_3 \ , \qquad (1.251)$$

where the *density–of–current matrix* is defined by

$$\vec{j}\,(x_1, x_2, t) := \frac{\vec{\nabla}_1 - \vec{\nabla}_2}{2mi}\rho_1(x_1, x_2, t) \ . \qquad (1.252)$$

The relation between (1.252) and the density–of–current operator (1.238) reads

$$\langle \hat{\vec{j}}\,(x, t)\rangle = \lim_{x' \to x}\vec{j}\,(x, x', t) \ .$$

The evolution equation for the matrix in (1.252) is

$$i\frac{\partial}{\partial t}\vec{j}\,(x_1, x_2, t) = [K_1(x_1, t) - K_1(x_2, t)]\,\vec{j}\,(x_1, x_2, t)-$$

$$-\frac{i}{2m}\left[\vec{\nabla}_1 U(x_1, t) + \vec{\nabla}_2 U(x_2, t)\right]\rho_1(x_1, x_2, t)-$$

$$-\frac{i}{2m}\int \left[\vec{\nabla}_1 \Phi(x_1, x_3) + \vec{\nabla}_2 \Phi(x_2, x_3)\right]\rho_2(x_3, x_1, x_3, x_2, t)dx_3-$$

$$-\frac{i}{2m}\int\left[\Phi(x_1,x_3)-\Phi(x_2,x_3)\right]\left(\vec{\nabla}_1-\vec{\nabla}_2\right)\rho_2(x_3,x_1,x_3,x_2,t)dx_3 , \qquad (1.253)$$

where the 2–matrix satisfies (1.249).

The evolution equations for matrices are to be supplemented with initial and boundary conditions. However, the system of these equations is not closed, since the equation for $\rho_1$ or $\vec{j}$ involves $\rho_2$, the equation for $\rho_2$ involves $\rho_3$. Generally, an equation for $\rho_n$ always contains $\rho_{n+1}$, of course, if the interaction potential is not zero. In this way, reduced density matrices satisfy an infinite system of integrodifferential equations, which is called the *hierarchical chain of equations*. To obtain a closed set of equations one has to break this chain by approximately expressing one of $\rho_n$ through the lower–order density matrices. Then one gets a closed system of $n-1$ equations for $\rho_1,\rho_2,\ldots,\rho_{n-1}$. This kind of approximation is called the *decoupling procedure*. It is important that any approximation used would preserve the general properties of density matrices following from their definition (1.96). This is the conjugation property

$$\rho_n^*(x^n,y^n,t)=\rho_n(y^n,x^n,t); \qquad (1.254)$$

the symmetry property

$$P(x_i,x_j)\rho_n(x^n,y^n,t)=\pm\rho_n(y^n,x^n,t) , \qquad (1.255)$$

where $x_i,x_j\in x^n$; the non–negativeness of the diagonal elements

$$\rho_n(x^n,x^n)\geq 0; \qquad (1.256)$$

and the normalization condition (1.103). For the density–of–current matrix, given by (1.252), in compliance with (1.254), one has

$$\vec{j}^{\,*}(x_1,x_2,t)=\vec{j}(x_2,x_1,t) .$$

Note that in some cases the symmetry property (1.255) can be broken giving, nevertheless, a good approximation [43-46]. For example, the widely used *Hartree approximation*

$$\rho_2(x_1,x_2,y_1,y_2,t)\cong\rho_1(x_1,y_1,t)\rho_1(x_2,y_2,t) , \qquad (1.257)$$

obviously does not obey (1.255). The property (1.255) holds for the *Hartree–Fock approximation*

$$\rho_2(x_1,x_2,y_1,y_2,t)\cong\rho_1(x_1,y_1,t)\rho_1(x_2,y_2,t)\pm\rho_1(x_1,y_2,t)\rho_1(x_2,y_1,t) . \qquad (1.258)$$

Substituting a decoupling, like (1.257) or (1.258), into (1.248), one obtains a closed equation for 1–matrix,

$$i\frac{\partial}{\partial t}\rho_1(x_1,x_2,t)=\left[K_1(x_1,t)-K_1(x_2,t)\right]\rho_1(x_1,x_2,t)+$$

$$+ \int \left[ \Phi(x_1, x_3) - \Phi(x_2, x_3) \right] \left[ \rho_1(x_1, x_2, t)\rho_1(x_3, x_3, t) + \right.$$

$$\left. + \varepsilon \rho_1(x_1, x_3, t)\rho_1(x_3, x_2, t) \right] dx_3 , \tag{1.259}$$

in which $\varepsilon = 0$ corresponds to the Hartree approximation and $\varepsilon = \pm 1$, to the Hartree–Fock approximation. Initial and boundary conditions are supposed to be given for equation (1.259) whose solution has also to satisfy the conditions

$$\rho_1^*(x_1, x_2, t) = \rho_1(x_2, x_1, t) , \qquad \rho_1(x, x, t) \geq 0 ,$$

$$\frac{1}{V} \int \rho_1(x, x, t)dx = \rho , \tag{1.260}$$

where $\rho$ is an average density of particles.

Actually, equation (1.259) is a set of two equations, since 1–matrix is a complex function

$$\rho_1(x_1, x_2, t) := a(x_1, x_2, t) + ib(x_1, x_2, t); \tag{1.261}$$

here $a$ and $b$ are, respectively, the real and imaginary parts of $\rho_1$. Owing to (1.260), we have

$$a(x_2, x_1, t) = a(x_1, x_2, t), \qquad b(x_2, x_1, t) = -b(x_1, x_2, t),$$

$$a(x, x, t) \geq 0, \qquad b(x, x, t) = 0, \qquad \frac{1}{V} \int a(x, x, t)dx = \rho .$$

Then, from (1.259) and (1.261) we get

$$\frac{\partial}{\partial t}a(x_1, x_2, t) = \left[ K_1(x_1, t) - K_1(x_2, t) \right] b(x_1, x_2, t) +$$

$$+ \int \left[ \Phi(x_1, x_3) - \Phi(x_2, x_3) \right] \left\{ a(x_3, x_3, t)b(x_1, x_2, t) + \right.$$

$$\left. + \varepsilon \left[ a(x_1, x_3, t)b(x_3, x_2, t) + a(x_3, x_2, t)b(x_1, x_3, t) \right] \right\} dx_3 ,$$

$$\frac{\partial}{\partial t}b(x_1, x_2, t) = \left[ K_1(x_2, t) - K_1(x_1, t) \right] a(x_1, x_2, t) +$$

$$+ \int \left[ \Phi(x_2, x_3) - \Phi(x_1, x_3) \right] \left\{ a(x_1, x_2, t)a(x_3, x_3, t) + \right.$$

$$\left. + \varepsilon \left[ a(x_1, x_3, t)a(x_3, x_2, t) + b(x_1, x_3, t)b(x_3, x_2, t) \right] \right\} dx_3 .$$

These equations describe the dynamics of all, diagonal as well as off-diagonal, elements of 1–matrix.

In addition to the Hartree approximation (1.257) and Hartree–Fock approximation (1.258), another widely used decoupling is the *Hartree–Fock–Bogolubov approximation* [39]

$$\rho_2(x_1, x_2, y_1, y_2, t) \cong \rho_1(x_1, y_1, t)\rho_1(x_2, y_2, t) \pm$$

$$\pm \rho_1(x_1, y_2, t)\rho_1(x_2, y_1, t) + \langle \psi(x_1)\psi(x_2) \rangle \langle \psi^\dagger(y_2)\psi^\dagger(y_1) \rangle \tag{1.262}$$

containing the *anomalous averages* $\langle \psi\psi \rangle$ and $\langle \psi^\dagger\psi^\dagger \rangle$. In this case, to find a closed set of evolution equations for 1–matrix, one needs to write down the evolution equations for the anomalous averages. Group–theoretical analysis of the Hartree–Fock–Bogolubov approximation reveals a rich variety of admissible physical states [47].

## 1.10  Coherent States

The notion of coherent states appears in many applications, therefore it is worth presenting here the general mathematical definition and the main properties of these states [48-51].

*Coherent states* are microstates that are normalized eigenvectors of the annihilation operator. Taking the latter in the form (1.54), we have, by definition,

$$\hat{\psi}(x)h_\eta := \eta(x)h_\eta \,, \tag{1.263}$$

where $\eta(x)$ is a *coherent field*, and the coherent state

$$h_\eta = [\eta_n(x^n)] \in \mathcal{H} \tag{1.264}$$

pertains to the Hilbert space $\mathcal{H}$ given by (1.51).

The action of the annihilation operator (1.54) on the vector (1.264) yields

$$\hat{\psi}(x)h_\eta = \left[\sqrt{n+1}\,\eta_{n+1}(x^n, x)\right] \,.$$

Substituting this into (1.263), we have

$$\sqrt{n+1}\,\eta_{n+1}(x^n, x) = \eta(x)\eta_n(x^n) \,.$$

The solution to this recursion relation reads

$$\eta_n(x^n) = \frac{\eta_0}{\sqrt{n!}} \prod_{k=1}^{n} \eta(x_k) \,, \tag{1.265}$$

where $\eta_0 \in \mathbb{C}$ and $\eta(x)$ is an arbitrary complex function of a Hilbert space with the scalar product

$$(\eta, \xi) := \int \eta^*(x)\xi(x)dx \,.$$

The scalar product of coherent states is

$$h_\eta^+ h_\xi = \sum_{n=0}^{\infty} (\eta_n, \xi_n) = \sum_{n=0}^{\infty} \frac{\eta_0^* \xi_0}{n!}(\eta, \xi)^n \,,$$

which can be reduced to

$$h_\eta^+ h_\xi = \eta_0^* \xi_0 e^{(\eta, \xi)} \,.$$

From the normalization condition

$$h_\eta^+ h_\eta = 1$$

we find

$$|\eta_0| = \exp\left\{-\frac{1}{2}(\eta, \eta)\right\} \,. \tag{1.266}$$

Thus, coherent states are defined by (1.264), (1.265), and (1.266).

For the scalar product of coherent states, within a constant phase factor, we have

$$h_\eta^+ h_\xi = \exp\left\{-\frac{1}{2}(\eta,\eta) + (\eta,\xi) - \frac{1}{2}(\xi,\xi)\right\} , \qquad (1.267)$$

which shows that different states are not orthogonal to each other, since

$$0 < |h_\eta^+ h_\xi| \le 1 .$$

Introducing the corresponding functional integral with a differential measure $\mathcal{D}\eta$, one can get the resolution of unity

$$\int h_\eta h_\eta^+ \mathcal{D}\eta = \hat{1} . \qquad (1.268)$$

The formal expression (1.268) is to be interpreted in the sense of weak equality, that is, as equality of arbitrary matrix elements of the indicated expression.

Coherent states form a basis but it is overcomplete because of the absence of orthogonality. Also, coherent states are eigenvectors of only the annihilation operator (1.54), but not of the creation operator (1.55), whose action on (1.264) gives

$$\hat{\psi}^\dagger h_\eta = \left[\sqrt{n}\eta_{n-1}(x^{n-1})\delta(x_n - x)\right] .$$

However, there is a relation

$$h_\eta^+ \hat{\psi}^\dagger(x) = h_\eta^+ \eta^*(x) \qquad (1.269)$$

following from the Hermitian conjugation of (1.263).

Considering the double action

$$\hat{\psi}(x)\hat{\psi}(x')h_\eta = \eta(x)\eta(x')h_\eta ,$$

we see that, on the class of coherent states, annihilation operators commute,

$$\left[\hat{\psi}(x), \hat{\psi}(x')\right] h_\eta = 0 .$$

But there are no simple commutators involving creation operators, which is clear from the actions

$$\hat{\psi}^\dagger(x)\hat{\psi}^\dagger(x')h_\eta = \left[\sqrt{n(n-1)}\eta_{n-2}(x^{n-2})\delta(x_{n-1} - x')\delta(x_n - x)\right] ,$$

$$\hat{\psi}(x)\hat{\psi}^\dagger(x')h_\eta = [(n+1)\eta_n(x^n)\delta(x - x')] ,$$

$$\hat{\psi}^\dagger(x')\hat{\psi}(x)h_\eta = \left[n\eta_n(x^{n-1}, x)\delta(x_n - x')\right] .$$

Notice that function (1.265) is symmetric with respect to permutation of any two variables $x_i, x_j \in x^n$,

$$P(x_i, x_j)\eta_n(x^n) = \eta_n(x^n) .$$

Consequently, $\eta_n(x^n) \in \mathcal{H}_n^+$ and $h_\eta \in \mathcal{H}_+$, where $\mathcal{H}_n^+$ and $\mathcal{H}_+$ are defined in Section 1.3. In the same way as in (1.69), we may introduce the Bose field operators $\psi_+$ and

$\psi_+^\dagger$, which have, on the class of coherent states, simple commutation relations given in (1.71).

To introduce coherent states for the Fermi field operators, according to (1.69), requires that the function (1.265) would be antisymmetric. This is achievable if the coherent field $\eta(x)$ is not a complex function, as in the case of Bose statistics, but an element of the Grassmann algebra, satisfying the relation

$$\eta_F(x)\eta_F(x') + \eta_F(x')\eta_F(x) = 0 \, ,$$

in which the index $F$ refers to the Fermi case. If $\eta(x)$ is a *Grassmann variable*, then the function (1.265) is antisymmetric with respect to permutation of variables $x_i, x_j \in x^n$. Hence, $\eta_n(x^n) \in \mathcal{H}_n^-$, and $h_\eta \in \mathcal{H}_-$. Therefore, one can define the Fermi field operators $\psi_-$ and $\psi_+$, as in (1.69), which have, on the class of coherent states, the standard anticommutation relations, as in (1.71).

In this way, coherent states can be introduces as the normalized eigenvalues of Bose, as well as Fermi operators,

$$\psi(x)h_\eta := \eta(x)h_\eta \, , \tag{1.270}$$

where $\psi(x)$ is either $\psi_+$ or $\psi_-(x)$. The difference is that for Bose statistics $\eta(x)$ is a complex function, while for Fermi statistics $\eta(x)$ is an element of the Grassmann algebra. These two cases may be combined in one condition for the coherent fields

$$[\eta(x), \eta(x')]_\mp = 0 \, , \tag{1.271}$$

in which the upper sign, as usual, is related to Bose statistics, and the lower, to Fermi statistics.

Coherent states (1.264) can be obtained from vacuum by means of field operators. By analogy with (1.57) and (1.63), we may define *n–particle coherent states*

$$[\delta_{mn}\eta_n(x^n)] := \frac{1}{\sqrt{n!}} \int \eta_n(x^n) \prod_{k=1}^{n} \psi^\dagger(x_k)|0\rangle dx^n \, , \tag{1.272}$$

where $\eta(x)$ entering into (1.265), depending on the type of statistics considered, is either a complex function or a Grassmann variable, as is discussed above, and $\psi^\dagger(x)$ is, respectively, either a Bose creation operator, or Fermi creation operator. With the form (1.265), we can present (1.272) as

$$[\delta_{mn}\eta_n(x^n)] = \frac{\eta_0}{n!} \left( \int \psi^\dagger(x)\eta(x)dx \right)^n |0\rangle \, .$$

Then, for the coherent state (1.264) we get

$$h_\eta = \sum_{n=0}^{\infty} \frac{\eta_0}{n!} \left( \int \psi^\dagger(x)\eta(x)dx \right)^n |0\rangle \, .$$

As is easy to notice, the latter expression is a series expansion of an exponential

$$h_\eta = \eta_0 \exp\left\{ \int \psi^\dagger(x)\eta(x)dx \right\} |0\rangle \, .$$

Taking onto account the normalization (1.266), we come to

$$h_\eta = \exp\left\{-\frac{1}{2}(\eta,\eta) + \int \psi^\dagger(x)\eta(x)dx\right\}|0\rangle . \qquad (1.273)$$

This is a general formula for constructing coherent states from vacuum.

To find the time evolution of coherent states, let us generalize the definition (1.263) to the form

$$\psi(x,t)h_\eta = \eta(x,t)h_\eta , \qquad (1.274)$$

in which $\psi(x,t)$ is given by (1.75). Whence, the *time–dependent coherent state* is

$$h_\eta(t) := U(t)h_\eta , \qquad (1.275)$$

for which

$$\psi(x,0)h_\eta(t) = \eta(x,t)h_\eta(t) .$$

The Hermitian conjugate to (1.274) is

$$h_\eta^+\psi^\dagger(x,t) = h_\eta^+\eta^*(x,t) .$$

The evolution equation for (1.275) follows from the equation

$$i\frac{d}{dt}U(t) = HU(t)$$

for the unitary operator, which yields

$$i\frac{d}{dt}h_\eta(t) = Hh_\eta(t) . \qquad (1.276)$$

The scalar product of time–dependent coherent states

$$h_\eta^+(t)h_\xi(t) = h_\eta^+h_\xi = \eta_0^*\xi_0 e^{(\eta,\xi)}$$

does not depend on time.

Coherent states permit us to write simple expressions for the matrix elements of the operators from the algebra of observables. Thus, for the number–of–particle operator (1.208), using

$$h_\eta^+\psi^\dagger(x,t)\psi(x',t)h_\xi = \eta_0^*\xi_0 e^{(\eta,\xi)}\eta^*(x,t)\xi(x',t),$$

we get

$$h_\eta^+\hat N(t)h_\xi = \eta_0^*\xi_0 e^{(\eta,\xi)}\int \eta^*(x,t)\xi(x,t)dx .$$

The diagonal elements of this is

$$h_\eta^+\hat N(t)h_\eta = \int |\eta^*(x,t)|^2\, dx , \qquad (1.277)$$

where the normalization (1.266) is invoked. For the Hamiltonian (1.226), the diagonal
element reads

$$h_\eta^+ H(t) h_\eta = \int \eta^*(x,t) \left[ K(\vec{\nabla}) + U(x,t) - \mu \right] \eta(x,t) dx +$$

$$+ \frac{1}{2} \int |\eta(x,t)|^2 V(x,x') |\eta(x',t)|^2 dx dx' . \tag{1.278}$$

The evolution equation for a coherent field $\eta(x,t)$ can be derived from equation
(1.234) for the field operator and the use of the equality

$$h_\eta^+ i \frac{\partial}{\partial t} \psi(x,t) h_\eta = i \frac{\partial}{\partial t} \eta(x,t) . \tag{1.279}$$

In this way we obtain the *coherent–field equation*

$$\left\{ K(\vec{\nabla}) + U(x,t) - \mu + \int \Phi(x,x') |\eta(x',t)|^2 dx' \right\} \eta(x,t) = i \frac{\partial}{\partial t} \eta(x,t) . \tag{1.280}$$

The latter has the structure of a Schrödinger equation with an *effective coherent potential*

$$U_{eff}(x,t) := U(x,t) - \mu + \int \Phi(x,x') |\eta(x',t)|^2 dx' .$$

Since the part of (1.280), containing interparticle interactions is nonlinear with respect
to the coherent field $\eta(x,t)$, the equation (1.280) is a *nonlinear Schrödinger equation*.
This equation defines the coherent field up to a phase factor.

It is important to notice that nontrivial coherent states exist not for any interaction
potential, but only for those that are integrable, so that

$$\left| \int \Phi(x,x') dx' \right| < \infty .$$

If it is not so, then the corresponding term in (1.280) diverges and the coherent–field
equation has only a trivial solution $\eta(x,t) \equiv 0$.

# 2 Green's Functions

The method of Green's functions is one of the most powerful techniques in quantum statistical physics [52–57] and quantum chemistry [58].

If $L$ is a linear operator, and $f(x)$ is a given function, then a solution of $Lu = f$ can be expressed in the form $u(x) = \int G(x, \xi) f(\xi) d\xi$, where $G$ satisfies the equation $LG(x, \xi) = \delta(x - \xi)$. The function $G(1, 2)$ which serves as the kernel of the above integral operator, is called a *Green's Function*. This, because it was George Green $(1793 - 1841)$ who introduced these functions into the study of linear equations. Certain Green's functions are also called *propagators* because as follows from the above integral we can think of $G$ as transfering an $L-modulated$ effect of a "charge" $f(\xi)$ at $\xi$ to the point $x$.

In recent years, the use of Green's functions has become widespread in theoretical physics and especially in the study of the $N-$body problem in quantum mechanics. As we shall see below they occur in many different forms. A basic mode of classifying them is as *one*–particle, *two*–particle, ..., and $n$–particle propagators. A reduced density operator can be defined as an initial value of a corresponding type of a Green's function and so the latter are also restricted by the necessity of being $N-representable$. This fact was often neglected by early Green's function enthusiasts. For a statistical system a hierarchy of equations, involving $n-$ and $(n+1)$-particle Green's function from $n = 0$ to $n = N$, must be satisfied. Only thus can $N-$representability of the Green's functions be *guaranteed*. For large $N$ this is manifestly impossible, so in practice the hierarchy of equations is usually truncated by an *ansatz* at the level of the first, second or third order Green's functions. But this always leaves some doubt as to the extent of possible errors introduced by the *ansatz*.

An even more difficult obstacle in discussing the $N-$body problem is that it cannot be solved exactly if $N > 1$! So to make any progress, some form of approximation must be adopted. For 2 or 3 particles, the approach which Hylleraas used to explain the spectrum of atomic helium can sometimes give very accurate results. However, for larger $N$ one normally starts from a tractable initial approximation attempting to improve this by successive perturbations. During recent decades physicists have developed increasingly sophisticated methods for obtaining approximations to Green's functions, and for the application of perturbation techniques to them.

It has been our aim in this chapter to provide the reader with a comprehensive overview of the extensive resulting literature about Green's functions. We have not followed any one previous treatment. Rather, we have sought to gather together what we regard as the most important and most generally applicable aspects of the Green's function approach to the quantum mechanical $N-$body problem. It may be that, for some readers, our condensed treament will bring to mind the well–known canard: "in order to do theoretical physics all you need is a knowledge of perturbation theory and a mastery of the Greek alphabet" and they will feel submerged under formalism! If so, we apologize, and only plead that we were unable, otherwise, to present in reasonable length what we consider to be the state–of–the–art for Green's functions.

## 2.1  Principal Definitions.

We shall adopt the widespread convention of using Planck's constant divided by $2\pi$ and Boltzmann's constant as units; so $h/2\pi = 1$, $k_B = 1$.

In order to have a compact notation we shall often denote the particle coordinates $(x_i, t_i)$ by the index $i$, so that a function of $n$ particles may be written as

$$f(12\ldots n) := f(x_1, t_1; x_2, t_2; \ldots; x_n, t_n),$$

and a differential for $n$ particles becomes

$$d(12\ldots n) := dx_1 dt_1 dx_2 dt_2 \ldots dx_n dt_n.$$

Thus, for the field operator we shall often use

$$\psi(1) := \psi(x_1, t_1).$$

The distinction between Bose and Fermi statistics is basic to quantum mechanics and is encoded in the field operator. We have adopted the convention that when the symbols $\pm$ or $\mp$ appear in our formulas *the upper sign refers to bosons.*

It is a basic metaphysical assumption of most of us that *cause* precedes *effect* in time. In consequence a so–called *chronological* or *time–ordering operator*, $\hat{T}$, acting on field operators plays an important role in the theory of Green's functions.

$$\hat{T}\psi(1)\psi^{\dagger}(2) := \begin{cases} \psi(1)\psi^{\dagger}(2), & t_{12} > 0 \\ \pm\psi^{\dagger}(2)\psi(1), & t_{12} < 0 \end{cases}, \tag{2.1}$$

where

$$t_{12} := t_1 - t_2 = -t_{21}. \tag{2.2}$$

Similarly, when $\hat{T}$ acts on the products $\psi(1)\psi(2)$ and $\psi^{\dagger}(1)\psi^{\dagger}(2)$, it arranges the order of the field operators so that time increases from *right* to *left*. If a permutation of the operators is neccessary, the product is multiplied by $+1$ or $-1$ according as Bose or Fermi statistics is relevant.

Invoking the unit-step function:

$$\Theta(t) := \begin{cases} 0, & t < 0 \\ 1, & t > 0 \end{cases},$$

the time–ordered product (2.1) can be written as

$$\hat{T}\psi(1)\psi^{\dagger}(2) = \Theta(t_{12})\,\psi(1)\,\psi^{\dagger}(2) \pm \Theta(-t_{12})\,\psi^{\dagger}(2)\,\psi(1).$$

The time–ordered products of any number of field operators can be defined in a similar manner.

We now define five different types of Green's functions commonly used in many-body theory:

- *causal Green's function or propagator*

$$G\left(12\right) := -i < \hat{T}\psi\left(1\right)\psi^{\dagger}\left(2\right) >; \tag{2.3}$$

- *retarded Green's function*

$$G_{ret}(12) := -i\Theta\left(t_{12}\right) < \psi\left(1\right)\psi^{\dagger}\left(2\right) >; \tag{2.4}$$

- *advanced Green's function*

$$G_{adv}\left(12\right) := -i\Theta\left(t_{21}\right) < \psi^{\dagger}\left(2\right)\psi\left(1\right) >; \tag{2.5}$$

- *commutator and anticommutator retarded Green's function*

$$G^{\mp}_{ret} := -i\Theta\left(t_{12}\right) < \left[\psi\left(1\right),\psi^{\dagger}\left(2\right)\right]_{\mp} >; \tag{2.6}$$

- *commutator and anticommutator advanced Green's function*

$$G^{\mp}_{adv}\left(12\right) := -i\Theta\left(t_{21}\right) < \left[\psi\left(1\right),\psi^{\dagger}\left(2\right)\right]_{\mp} > . \tag{2.7}$$

Note that the average $< \cdot >$ may , depending on the context, be with respect to a pure state or with respect to an ensemble which may be temperature dependent and which may be in equilibrium or not.

The properties of these different Green's functions are similar to one another and there are fairly obvious relations among them so that in most situations it is as convenient to work with one of them as another. However, in our opinion the *causal Green's function* has a slight edge for the following reasons:

1. As a linear combination of the retarded and advanced Green's functions, the causal Green's function carries more information than either of those separately:

$$G\left(12\right) = G_{ret}\left(12\right) \pm G_{adv}\left(12\right); \tag{2.8}$$

2. Dealing with the causal Green's function is rather simpler than with the commutator or anticommutator Green's functions;

3. The causal Green's function, as we have defined it is accepted and used, almost universally, in both quantum statistical mechanics and quantum field theory.

Notice, further, that in the usual set–up in Hilbert space, operators act on the left of the symbols denoting states of the system. Thus when the operator is a product we interpret its action to mean that the factor on the right acts first. The chronological operator, $\hat{T}$, was defined with this in mind ensuring that the factor which acts first on the wave vector is at the earlier time. This corresponds to our intuitive feeling about causality, justifying the term "causal" applied to (2.3)

It turns out that $G(12)$ is closely associated with the time-dependent 1-matrix. In fact,

$$\rho_1(x_1, x_2, t) = \pm i \lim_{t_i \to t} G(12), \qquad (2.9)$$

where the limit is taken subject to $t_2 > t_1$.

Recall that here and elsewhere when $\pm$ or $\mp$ occur indicating Fermi or Bose stastistics, the *upper* sign refers to Bose statistics.

The diagonal elements of the 1–matrix give the value of particle density and are therefore positive, that is

$$\rho_1(x, x, t) = \rho(x, t).$$

From this we obtain the following condition on $G(12)$,

$$\pm i \lim_{t_i \to t} \lim_{x_2 \to x_1} G(12) \geq 0, \qquad (2.10)$$

where the limit is taken subject to the restriction $t_2 > t_1$.

All of the Green's functions defined above are one–particle functions. In an analogous manner we may define a two–particle causal Green's function, or *two–particle propagator*:

$$G_2(1234) := - < \hat{T}\psi(1)\psi(2)\psi^\dagger(3)\psi^\dagger(4) > . \qquad (2.11)$$

The definition (2.1) for the chronological operator is generalized to an arbitrary finite product of field operators as follows: the factors are permuted so that time increases from right to left and the appropriate sign is determined by decomposing the required permutation into a succession of transpositions of neighbouring factors. It then follows that (2.11) has the following symmetry properties,

$$G_2(1234) = \pm G_2(2134) = \pm G_2(1243).$$

We again find that there is a simple relation between the 2–particle propagator and the time–dependent 2–matrix, namely

$$\rho_2(x_1, x_2, x_3, x_4, t) = \mp \lim_{t_i \to t} G_2(1234), \qquad (2.12)$$

where the limit is taken subject to the restriction $t_4 > t_3 > t_1 > t_2$.

Similarly, we define the *n–particle propagator* as

$$G_n(12\ldots 2n) := (-i)^n < \hat{T}\psi(1)\ldots\psi(n)\psi^\dagger(n+1)\ldots\psi^\dagger(2n) > . \qquad (2.13)$$

Again, we can recover the $n$–matrix from (2.13). It turns out that

$$\rho_n(x_1, \ldots, x_{2n}, t) = (\pm 1)^{(3n-1)n/2} i^n \lim_{t_i \to t} G_n(1\ldots 2n), \qquad (2.14)$$

where the limit is subject to the restriction

$$t_{2n} > t_{2n-1} > \ldots > t_{n+1} > t_1 > t_2 > \ldots > t_n.$$

In all of the Green's functions defined so far there are the same number of creation and annihilation operators. This is appropriate for situations in which the number of particles is conserved for in such cases the expected value for operators involving unequal numbers of creation an annihilation operators will be zero. However, if the number of particles is not conserved then such averages may be different from zero. Indeed, this can happen for systems with a Bose condensate and for superconductors. In these cases averages of operators such as $\psi(1)\psi(2)$, $\psi^\dagger(1)\psi^\dagger(2)$, $\psi^\dagger(1)\psi(2)\psi(3)$ may differ from zero. Such averages, which can differ from zero only if the number of particles is not conserved are often called *anomalous averages* and the Green's functions defined by means of such averages are called *anomalous Green's functions*. The most commonly used are certainly

$$F(12) := -i < \hat{T}\psi(1)\psi(2) >,$$

$$F^+(12) := i < \hat{T}\psi^\dagger(2)\psi^\dagger(1) > . \tag{2.15}$$

In analogy with the definition of Green's functions by means of the field operators, it is possible to employ any other operators to define Green's functions as statistical averages of products of operators. All observable quantities can be expressed by means of integrals involving Green's functions. For example it follows from (2.9) that the density of particles is

$$\rho(x,t) = \pm \lim_{x_1 \to x} \lim_{(21)} G(12). \tag{2.16}$$

Here we have employed an abbreviated notation for the limit to mean

$$\lim_{(21)} \Leftrightarrow x_2 \to x_1; \qquad t_i \to t, \qquad (t_2 > t_1). \tag{2.17}$$

Further, the density of particle current can be expressed as

$$j(x,t) = \pm \frac{1}{2m} \lim_{x_1 \to x} \lim_{(21)} (\nabla_1 - \nabla_2) G(12) \tag{2.18}$$

where the limit is interpreted by means of (2.17).

Any 1–particle operator, $\hat{A}_1(t)$, can be expressed by means of the field operators and an operator-valued function $A_1(x,t)$ as

$$\hat{A}_1(t) = \int \psi^\dagger(x,t) A_1(x,t) \psi(x,t) dx,$$

and its *average* or *expected value* can be written as

$$< \hat{A}_1 >= \pm i \int \lim_{(21)} A_1(x_1,t) G(12) dx_1 . \tag{2.19}$$

Similarly, the average, or expected value, of a 2–particle operator

$$\hat{A}_2(t) = \frac{1}{2} \int \psi^\dagger(x_1,t)\psi^\dagger(x_2,t) A_2(x_1,x_2,t)\psi(x_2,t)\psi(x_1,t) dx_1 dx_2$$

is

$$< \hat{A}_2 >= -\frac{1}{2} \int \lim_{(4321)} A_2(x_1, x_2, t) G_2(1234) dx_1 dx_2, \qquad (2.20)$$

where in this limit:

$$x_4 \to x_1, \qquad x_3 \to x_2; \qquad t_i \to t, \qquad (t_4 > t_3 > t_2 > t_1). \qquad (2.21)$$

In order to abbreviate notation, we shall use $x^n$, with $n$ as a *superscript*, to denote the $n$ variables $(x_1, x_2, \ldots, x_n)$ and thus $dx^n := dx_1 dx_2 \ldots dx_n$. Then an $n$-particle operator can be written as

$$\hat{A}_n(t) = \frac{1}{n!} \int \psi^\dagger(x_1, t) \ldots \psi^\dagger(x_n, t) A_n(x^n, t) \psi(x_n, t) \ldots \psi(x_1, t) dx^n.$$

The average of this operator is given by the formula

$$< \hat{A}_n >= \frac{(\pm i)^n}{n!} \int \lim_{(2n \ldots 1)} A_n(x^n, t) G_n(1 \ldots 2n) dx^n \qquad (2.22)$$

which involves the $n$-particle propagator (2.13) and the generalization of the limiting procedures (2.17) and (2.21) defined by the rule

$$\lim_{(2n \ldots 1)} := \begin{cases} x_{2n-i} \to x_{i+1}, & 0 \le i \le n-1 \\ t_j \to t, & t_{j+1} > t_j, \ \forall j. \end{cases} \qquad (2.23)$$

Applying this to the number-of-particles operator, $\hat{N}$, we find that the average number of particles is

$$N =< \hat{N} >= \pm i \int \lim_{(21)} G(12) dx_1. \qquad (2.24)$$

Averaging the Hamiltonian, $H$, we find

$$< H >= \pm i \int \lim_{(21)} [K(\nabla_1) + U(x_1, t) - \mu] G(12) dx_1 -$$

$$-\frac{1}{2} \int \lim_{(4321)} V(x_1, x_2) G_2(1234) dx_1 dx_2. \qquad (2.25)$$

Using this we obtain the internal energy as

$$E =< H > +\mu N. \qquad (2.26)$$

These examples should convince the reader that an *exact* knowledge of the Green's functions for an $N$-body system contains complete information about its state whether or not it is in a pure or statistical state. However, we also know that for $N > 1$ it is too much to hope for *exactness*. So our problem is to develop methods which can give us reasonable approximations.

## 2.2 Spectral Representations

The Green's functions defined in the previous section are appropriate whether or not the system being studied is in equilibrium. When it is in equilibrium, as suggested by Lehmann, taking the Fourier transform with respect to the time variable leads to a form for the Green's functions which is referred to as the *spectral representation* and with which it is much easier to work. The properties of this representation give rise to certain advantages of the Green's function method. In particular, singularities in the spectral representation of the first order Green's function occur at energy levels of quasi–particles.

For the equilibrium states we can introduce the *equilibrium correlation functions:*

$$C_+(x_1, x_2, t) := < \psi(1)\psi^\dagger(2) >,$$

$$C_-(x_1, x_2, t) := < \psi^\dagger(2)\psi(1) >, \tag{2.27}$$

in which

$$t = t_{12} = t_1 - t_2. \tag{2.28}$$

The right hand side of the equations in (2.27) apparently depends on $t_1$ and $t_2$ but, in fact, only on the difference $t$. For recall that when the Hamiltonian, $H$ , does not depend explicitly on $t$, i.e. $\partial H / \partial t = 0$, the evolution operator is given by

$$U(t) = e^{-iHt}; \qquad [U(t), \hat{\rho}] = 0,$$

where $\hat{\rho}$ is the $N$–particle von Neumann density matrix of the system. By definition,

$$< \psi(1)\psi^\dagger(2) >= Tr\left[\hat{\rho}\psi(1)\psi^\dagger(2)\right] =< \psi(x_1, t)\psi^\dagger(x_2, 0) >$$

since $\psi(x, t) = U^+(t)\psi(x, 0)U(t)$

The functions (2.27) enjoy the following properties with respect to complex conjugation.

$$C_+^*(x_1, x_2, t) = C_+(x_2, x_1, t) \qquad C_-^*(x_1, x_2, t) = C_-(x_2, x_1, t). \tag{2.29}$$

Recall that $\beta$ denotes the inverse temperature. An important relation between the correlation functions (2.27) is the *equilibrium condition:*

$$C_-(x_1, x_2, t) = C_+(x_1, x_2, t - i\beta), \tag{2.30}$$

which is often referred to as the *Kubo–Martin–Schwinger conditon*. Its proof follows:

$$C_-(x_1, x_2, t) = Tr\left[\hat{\rho}\psi^\dagger(x_2, 0)\psi(x_1, t)\right] = Tr\left[\psi(x_1, t)\hat{\rho}\psi^\dagger(x_2, 0)\right] =$$

$$= Tr\left[\hat{\rho}e^{\beta H}\psi(x_1, t)e^{-\beta H}\psi^\dagger(x_2, 0)\right] = Tr[\hat{\rho}\psi(x_1, t - i\beta)\psi^\dagger(x_2, 0)] = C_+(x_1, x_2, t - i\beta)$$

since, at thermal equilibrium, $\hat{\rho} = (\exp(-\beta H))/Tr(\exp(-\beta H))$.

The Fourier transforms, $K_\pm$, of functions (2.27) are given by

$$C_+(x_1, x_2, t) = \int_{-\infty}^{+\infty} K_+(x_1, x_2, \omega)e^{-i\omega t}\frac{d\omega}{2\pi}\,,$$

$$K_+(x_1, x_2, \omega) = \int_{-\infty}^{+\infty} C_+(x_1, x_2, t)e^{i\omega t}dt\,,,$$

$$C_-(x_1, x_2, t) = \int_{-\infty}^{+\infty} K_-(x_1, x_2, \omega)e^{-i\omega t}\frac{d\omega}{2\pi}\,,$$

$$K_-(x_1, x_2, \omega) = \int_{-\infty}^{+\infty} C_-(x_1, x_2, t)e^{i\omega t}dt\,.$$

Then, taking the Fourier transform of (2.30) we obtain

$$K_-(x_1, x_2, \omega) = e^{-\beta\omega}K_+(x_1, x_2, \omega)\,. \tag{2.31}$$

The correlation functions (2.27) at $t = 0$ satisfy the equation

$$C_+(x_1, x_2, 0) \mp C_-(x_1, x_2, 0) = \delta(x_1 - x_2)\,, \tag{2.32}$$

which follows from the commutation relations for the field operators. The Fourier transform of (2.32) reads

$$\int_{-\infty}^{+\infty} K_+(x_1, x_2, \omega)\left(1 \mp e^{-\beta\omega}\right)\frac{d\omega}{2\pi} = \delta(x_1 - x_2)\,, \tag{2.33}$$

when we take (2.31) into account.

By means of the correlation functions (2.27) it is possible to rewrite the Green's functions of (2.4) and (2.5) as

$$G_{ret}(12) = -i\Theta(t)C_+(x_1, x_2, t), \qquad G_{adv}(12) = -i\Theta(-t)C_-(x_1, x_2, t), \tag{2.34}$$

where the notation (2.28) is used. The Fourier transforms, $G_+$ and $G_-$, of these Green's functions are defined by the relations:

$$G_{ret}(12) = \frac{1}{2\pi}\int_{-\infty}^{+\infty} G_+(x_1, x_2, \omega)e^{-i\omega t}d\omega,$$

$$G_{adv}(12) = \frac{1}{2\pi}\int_{-\infty}^{+\infty} G_-(x_1, x_2, \omega)e^{-i\omega t}d\omega. \tag{2.35}$$

The unit step function, $\Theta$, can be expressed as a Fourier integral as follows:

$$\Theta(\pm t) = \pm i\int_{-\infty}^{+\infty} \frac{e^{-i\omega t}}{\omega \pm i0}\frac{d\omega}{2\pi}, \tag{2.36}$$

where the convention

$$\int_{-\infty}^{+\infty} f(\omega \pm i0)d\omega := \lim_{\varepsilon\to +0}\int_{-\infty}^{+\infty} f(\omega \pm i\epsilon)\,d\omega \tag{2.37}$$

is used.

Recall the well–known formulas for the $\delta$–function:

$$\delta(t) = \int_{-\infty}^{+\infty} e^{-i\omega t}\frac{d\omega}{2\pi} \qquad \int_{-\infty}^{+\infty} e^{i\omega t}dt = 2\pi\,\delta(\omega).$$

To obtain expressions for $G_+$ and $G_-$, we invert equations (2.35) making use of (2.34) and the convolution theorem for Fourier integrals. This leads us to the *spectral representation* of the retarded and of the advanced Green's functions

$$G_+(x_1, x_2, \omega) = -\frac{1}{2\pi}\int_{-\infty}^{+\infty}\frac{K_+(x_1, x_2, \omega')}{\omega' - \omega - i0}d\omega',$$

$$G_-(x_1, x_2, \omega) = \frac{1}{2\pi}\int_{-\infty}^{+\infty}\frac{K_-(x_1, x_2, \omega')}{\omega' - \omega + i0}d\omega'. \tag{2.38}$$

The spectral representation of the causal Green's function is an immediate consequence of (2.38) and the definition of the Fourier transform of (2.3) as

$$G(x_1, x_2, \omega) = \int_{-\infty}^{+\infty} G(12)e^{i\omega t}dt, \qquad G(12) = \int_{-\infty}^{+\infty} G(x_1, x_2, \omega)e^{-i\omega t}\frac{d\omega}{2\pi}. \tag{2.39}$$

According to (2.8) we have

$$G(x_1, x_2, \omega) = G_+(x_1, x_2, \omega) \pm G_-(x_1, x_2, \omega). \tag{2.40}$$

Combining (2.38) and (2.40), we obtain the *spectral representation*

$$G(x_1, x_2, \omega) = -\int_{-\infty}^{+\infty} K_+(x_1, x_2, \omega')\left(\frac{1}{\omega' - \omega - i0} \mp \frac{e^{-i\beta\omega'}}{\omega' - \omega + i0}\right)\frac{d\omega'}{2\pi}, \tag{2.41}$$

of the causal Green's function.

For systems in equilibrium, the averages of operators are expressed by means of integrals involving $G(x_1, x_2, \omega)$ using the formula (2.41). For example, the density of particles (2.16) is

$$\rho(x, t) = \pm i \lim_{t\to+0}\int_{-\infty}^{+\infty} e^{i\omega t}G(x, x, \omega)\frac{d\omega}{2\pi}. \tag{2.42}$$

The Green's function at coincident times is defined as

$$G(12)\,|_{t=0} := \lim_{t\to-0} G(12). \tag{2.43}$$

Substituting (2.41) into (2.42), we obtain

$$\rho(x, t) = \int_{-\infty}^{+\infty} K_+(x, x, \omega)\,e^{-\beta\omega}\frac{d\omega}{2\pi}.$$

Instead of dealing with the two Fourier transforms $K_+$ and $K_-$ which are related by (2.31), we introduce the *spectral function*

$$J(x_1, x_2, \omega) := \frac{K_+(x_1, x_2, \omega)}{1 \pm n(\omega)} = \frac{K_-(x_1, x_2, \omega)}{n(\omega)}. \tag{2.44}$$

Here,

$$n(\omega) := \frac{1}{e^{\beta\omega} \mp 1} \tag{2.45}$$

is called the *Bose function* or the *Fermi function* according as one takes the minus or plus sign. Recall that this is consistent with our general convention that the *upper* of the two signs refers to Bose statistics, the *lower* to Fermi statistics.

Using the equality $1 \pm n(\omega) = n(\omega)e^{\beta\omega}$ and the spectral function (2.44) the spectral representation of the causal Green's function, or propagator, can be expressed in the form

$$G(x_1, x_2, \omega) = -\int_{-\infty}^{+\infty} J(x_1, x_2, \omega') \left[ \frac{1 \pm n(\omega')}{\omega' - \omega - i0} \mp \frac{n(\omega')}{\omega' - \omega + i0} \right] \frac{d\omega'}{2\pi}. \tag{2.46}$$

Expressions involving poles can be reorganized by invoking the symbolic identities, introduced by Dirac, in which the symbol $P$ indicates that the *principal* value is to be taken.

$$\frac{1}{\omega \pm i0} := P\frac{1}{\omega} \mp i\pi\,\delta(\omega),$$

$$\frac{1}{f(\omega \pm i0)} := P\frac{1}{f(\omega)} \mp i\pi\,\delta(f(\omega)). \tag{2.47}$$

Evaluating (2.46) by means of (2.47), we find that

$$G(x_1, x_2, \omega) = P\int_{-\infty}^{+\infty} \frac{J(x_1, x_2, \omega')}{\omega - \omega'}\frac{d\omega'}{2\pi} - \frac{i}{2}\left[1 \pm 2n(\omega)\right] J(x_1, x_2, \omega). \tag{2.48}$$

If $J$ has no poles on the real axis then again using (2.47) we can show that (2.48) implies that

$$J(x_1, x_2, \omega) = i\left[G(x_1, x_2, \omega + i0) - G(x_1, x_2, \omega - i0)\right]. \tag{2.49}$$

## 2.3   Dispersion Relations

As follows from their spectral representations (2.38) the Fourier transforms of the retarded and advanced Green's functions, $G_+$ and $G_-$, have the following asymptotic behaviour, when $\omega \longrightarrow \pm\infty$

$$G_+(x_1, x_2, \omega) \cong \frac{1}{\omega}C_+(x_1, x_2, 0),$$

and

$$G_-(x_1, x_2, \omega) \cong -\frac{1}{\omega}C_-(x_1, x_2, 0).$$

Thus

$$\lim_{\omega \to \pm\infty} G_+(x_1, x_2, \omega) = \lim_{\omega \to \pm\infty} G_-(x_1, x_2, \omega) = 0.$$

It follows from (2.34) that the retarded Green's function differs from zero only if $t > 0$, and the advanced Green's function only if $t < 0$. From this and the expression (2.35) for the Fourier transforms, we may conclude that $G_+$ can be continued analytically into the upper complex $\omega$ half–plane and $G_-$ into the lower half–plane. For $\omega$ in the domain of analyticity of $G_+$ or $G_-$ one has the Cauchy integrals,

$$G_+(x_1, x_2, \omega) = \frac{1}{2\pi i} \int_{-\infty}^{+\infty} \frac{G_+(x_1, x_2, \omega')}{\omega' - \omega - i0} d\omega',$$

$$G_-(x_1, x_2, \omega) = -\frac{1}{2\pi i} \int_{-\infty}^{+\infty} \frac{G_-(x_1, x_2, \omega')}{\omega' - \omega + i0} d\omega'. \tag{2.50}$$

Using the identities (2.47) and (2.50) we obtain the *Hilbert transforms*

$$\mathrm{Im}G_\pm(x_1, x_2, \omega) = \mp \frac{P}{\pi} \int_{-\infty}^{+\infty} \frac{\mathrm{Re}G_\pm(x_1, x_2, \omega')}{\omega' - \omega} d\omega',$$

$$\mathrm{Re}G_\pm(x_1, x_2, \omega) = \pm \frac{P}{\pi} \int_{-\infty}^{+\infty} = \frac{\mathrm{Im}G_\pm(x_1, x_2, \omega')}{\omega' - \omega} d\omega'. \tag{2.51}$$

In analogy with similar equations in Optics, (2.51) are called the *dispersion relations* for the retarded and advanced Green's functions.

The Fourier transform of the causal Green's function has the asymptotic form

$$G(x_1, x_2, \omega) \cong \frac{1}{\omega} \delta(x_1 - x_2), \qquad (\omega \longrightarrow \pm\infty).$$

The propagator is a linear combination (2.40) of $G_+$ and $G_-$, but since these have non–intersecting domains of analyticity for complex $\omega$ there are no simple dispersion relations for the Fourier transform of the causal Green's function. Even so, there are dispersion relations for its diagonal element

$$G(x, \omega) := G(x, x, \omega). \tag{2.52}$$

To derive these we first note that it follows from (2.10) that

$$C_-(x, x, 0) \geq 0. \tag{2.53}$$

From this and (2.31) we obtain the inequality

$$\int_{-\infty}^{+\infty} K_+(x, x, \omega) e^{-\beta\omega} d\omega \geq 0,$$

valid for all $\beta \in \mathbb{R}_+$. Hence,

$$K(x, \omega) := K_+(x, x, \omega) \geq 0. \tag{2.54}$$

From the spectral representation (2.41), by means of (2.47), we obtain

$$\mathrm{Re}G(x, \omega) = \frac{P}{\pi} \int_{-\infty}^{+\infty} \frac{\mathrm{Im}G(x, \omega')}{\omega' - \omega} \left( \frac{e^{\beta\omega'} \mp 1}{e^{\beta\omega'} \pm 1} \right) d\omega',$$

$$\operatorname{Im}G(x,\omega) = -\frac{1}{2}K(x,\omega)\left(1 \pm e^{-\beta\omega}\right). \tag{2.55}$$

These are the *dispersion relations* for the causal Green's function or the propagator. Rewriting (2.55) separately for the Bose and Fermi cases, we obtain

$$\operatorname{Re}G(x,\omega) = \frac{P}{\pi}\int_{-\infty}^{+\infty}\frac{\operatorname{Im}G(x,\omega')}{\omega'-\omega}\tanh\left(\frac{\beta\omega'}{2}\right)d\omega', \qquad (Bose)$$

$$\operatorname{Re}G(x,\omega) = \frac{P}{\pi}\int_{-\infty}^{+\infty}\frac{\operatorname{Im}G(x,\omega')}{\omega'-\omega}\coth\left(\frac{\beta\omega'}{2}\right)d\omega'. \qquad (Fermi)$$

The equality (2.55) gives

$$K(x,\omega) = -\frac{2\operatorname{Im}G(x,\omega)}{1\pm e^{-\beta\omega}}.$$

Using this, we may write the density of particles for a system in equilibrium as

$$\rho(x,t) = -\frac{1}{\pi}\int_{-\infty}^{+\infty}\frac{\operatorname{Im}G(x,\omega)}{e^{\beta\omega}\pm 1}d\omega. \tag{2.56}$$

Recall that for equilibrium this density does not vary with time.

Another form of the dispersion relations for the causal Green's function can be expressed by means of the diagonal element of the spectral function (2.44) which we denote by

$$J(x,\omega) := J(x,x,\omega), \tag{2.57}$$

which, as follows from (2.54), is a real valued function. Then, from the spectral representation (2.46), we obtain the dispersion relations

$$\operatorname{Re}G(x,\omega) = -P\int_{-\infty}^{+\infty}\frac{J(x,\omega')}{\omega'-\omega}\frac{d\omega'}{2\pi},$$

$$\operatorname{Im}G(x,\omega) = -\frac{1}{2}[1\pm 2n(\omega)]J(x,\omega), \tag{2.58}$$

where the equality

$$1\pm 2n(\omega) = \frac{e^{\beta\omega}\pm 1}{e^{\beta\omega}\mp 1}$$

has been used.

If we make use of (2.57), the expression (2.56) for the particle density takes the simpler form

$$\rho(x,t) = \int_{-\infty}^{+\infty} J(x,\omega)n(\omega)\frac{d\omega}{2\pi}. \tag{2.59}$$

Other observables can similarly be expressed by means of integrals involving the spectral function (2.57).

For equilibrium systems there is still one more very important type of Green's function – the *temperature Green's functions*, $G_{tem}$, the definition of which is analogous to

64

that for the causal Green's function except that *imaginary* time variables are employed. Set

$$t_j = -i\tau_j, \qquad (j = 1, 2, \ldots)$$

with $\tau_j$ in the interval $0 \leq \tau_j \leq \beta$. The definition of the *one -particle temperature Green's function* then takes the form

$$G_{tem}(12) := -i\left[\Theta(\tau)C_+(x_1, x_2, -i\tau) \pm \Theta(-\tau)C_-(x_1, x_2, -i\tau)\right], \qquad (2.60)$$

where $\tau := \tau_1 - \tau_2 = i(t_1 - t_2)$.

The equilibrium condition (2.30) for the temperture Green's function then becomes:

$$G_{tem}(12)\,|_{\tau=0} = \pm G_{tem}(12)_{\tau=\beta}. \qquad (2.61)$$

Because of this periodicity condition, $G_{tem}$ may be expressed as a Fourier series. Spectral representations and dispersion relations for the temperature Green's function can be derived which are strictly analogous to those for the real–time propagator.

Which type of Green's function one wishes to use is largely a matter of taste. However, in our view, the real–time causal Green's functions are the most convenient and so, normally, we prefer to deal with them.

## 2.4    Evolution Equations

The previous two sections were devoted to the study of Green's functions for systems of identical particles in equilibrium. We now consider arbitrary statistical systems which may or may not be in equilibrium and obtain the equations of motion for the various Green's functions. These equations are practically the same for Advanced, Retarded and Causal Green's functions. For the sake of concreteness we shall work mostly with Causal Green's functions that is with *propagators*.

Assume that the system is specified by the Hamiltonian:

$$H(t) = \int \psi^\dagger(x, t)\left[K(\nabla) + U(x, t) - \mu\right]\psi(x, t)dx +$$

$$+\frac{1}{2}\int \psi^\dagger(x, t)\psi^\dagger(x', t)V(x, x')\psi(x', t)\psi(x, t)dxdx',$$

in which $K(\nabla)$ is a kinetic–energy operator. Introduce the *effective chemical potential*

$$\mu(1) := \mu - U(x_1, t_1), \qquad (2.62)$$

which incorporates the external fields in $U(1)$, and define the *interaction potential*

$$\Phi(12) := \Phi(x_1, x_2)\delta(t_1 - t_2), \qquad (2.63)$$

where $\Phi(x_1, x_2) := \frac{1}{2}[V(x_1, x_2) + V(x_2, x_1)]$.

In what follows we shall occasionally employ propagators with coincident variables. In order to avoid ambiguity, we generalize the definition of the chronological operator in

(2.1) to the case of two field operators with the same argument and adopt the following convention:

$$\hat{T}\psi(1)\psi^{\dagger}(1) := \pm\psi^{\dagger}(1)\psi(1) \qquad \hat{T}\psi^{\dagger}(1)\psi(1) := \psi^{\dagger}(1)\psi(1). \tag{2.64}$$

A product of two field operators in which the creation operator is to the left of the annihilator has been called the *normal form*. Thus, the chronological operator by acting on the product of $\psi(1)$ and $\psi^{\dagger}(1)$, reduces it to normal form. This language occurred first in connection with the well–known theorem of Wick .

The equation of motion for the one–particle propagators follows from the Heisenberg evolution equation for the field operator and the equality

$$\frac{d}{dt}\Theta\left(\pm t\right) = \pm\delta\left(t\right).$$

This implies

$$\left[i\frac{\partial}{\partial t_1} - K(\nabla_1) + \mu(1)\right]G(12) \mp i\int \Phi(13)G_2(1332)d(3) = \delta(12), \tag{2.65}$$

where

$$\delta(12) = \delta(x_1 - x_2)\delta(t_1 - t_2).$$

A quantity which plays an important role in the theory of Green's functions is the *self–energy*, $\Sigma(12)$, which is defined implicitly by the equation

$$\int \Sigma(13)G(32)d(3) = \pm\int \Phi(13)G_2(1332)d(3). \tag{2.66}$$

Then (2.65) can be transformed to the form

$$\left[i\frac{\partial}{\partial t} - K(\nabla_1) + \mu(1)\right]G(12) - \int \Sigma(13)G(32)d(3) = \delta(12). \tag{2.67}$$

Defining the *inverse propagator*

$$G^{-1}(12) := \left[i\frac{\partial}{\partial t} - K(\nabla_1) + \mu(1)\right]\delta(12) - \Sigma(12), \tag{2.68}$$

we may write (2.67) as

$$\int G^{-1}(13)G(32)d(3) = \delta(12). \tag{2.69}$$

It then follows from ( 2.66) that

$$\Sigma(12) = \pm i\int \Phi(13)G_2(1334)G^{-1}(42)d(34). \tag{2.70}$$

Suppose that $G_0(12)$ is any propagator for which there is an inverse, $G_0^{-1}(12)$, satisfying the equation

$$\int G_0^{-1}(13)G_0(32)d(3) = \delta(12), \tag{2.71}$$

then (2.69) can be transformed into the integral equation

$$G(12) = G_0(12) + \int B(13)G(32)d(3), \qquad (2.72)$$

where the kernel

$$B(12) := \int G_0(13)[G_0^{-1}(32) - G^{-1}(32)]d(3). \qquad (2.73)$$

The equation (2.72) is called the *Dyson equation* which is thus a transform, via (2.69), of the equation of motion (2.65).

If we choose

$$G_0^{-1}(12) := \left[i\frac{\partial}{\partial t_1} - K(\nabla_1) + \mu(1)\right]\delta(12) - \Sigma_0(12), \qquad (2.74)$$

then the kernel (2.73) becomes

$$B(12) = \int G_0(13)[\Sigma(32) - \Sigma_0(32)]d(3).$$

Since (2.70) and therefore $B(12)$ involve $G_2$, Dyson's equation does not by itself determine the one–particle Green's function $G(12)$. The equation of motion for $G_2$ involves $G_3$ and so on up to $G_N$. There is thus a hierarchy of equations for Green's functions as there is for Reduced Density Matrices. In order to short–circuit this chain of equations one makes an *ansatz* for the Green's function of some chosen order thus introducing approximations.

Perhaps the most common such an *ansatz* is to assume that

$$G_2(1234) = G_2^{HF}(1234) := G(14)G(23) \pm G(13)G(24), \qquad (2.75)$$

which is called the *Hartree–Fock approximation*. Making this ansatz will lead us to equations determing the 1–propagator.

The *Hartree–Fock ansatz* enables us to obtain the following expression for the *self–energy* (5.70),

$$\Sigma_{HF}(12) = \pm i\delta(12) \int \Phi(3)G(33)d(3) + i\Phi(12)G(12), \qquad (2.76)$$

where

$$G(11) := \lim_{x_2 \to x_1} \lim_{t_{12} \to -0} G(12), \qquad (2.77)$$

in accord with the convention (2.64). Introducing the *Hartree potential*

$$V_H(1) := \int \Phi(12)\rho(2)d(2), \qquad (2.78)$$

in which

$$\rho(1) = \pm iG(11) = \rho(x_1, t_1)$$

is the density of particles, we may reduce (2.76) to

$$\Sigma_{HF}(12) = V_H(1)\,\delta(12) + i\Phi(12)\,G(12).$$

There is a more general *ansatz* which appeared in connection with the theory of superconductivity and is now usually referred to as the *Hartree–Fock–Bogolubov approximation*. This implies that the two–particle propagator decouples as

$$G_2^{HFB}(1234) = G(14)G(23) \pm G(13)G(24) \mp F(12)F^+(34), \tag{2.79}$$

where $F$ and $F^+$ are the *anomalous Green's functions* defined in (2.15). This approximation is used not only for superconductors but also often for discussing Bose systems with condensate. Under assumption (2.79) the self–energy (2.70) takes the form

$$\Sigma_{HFB}(12) = \Sigma_{HF}(12) - i\int \Phi(13)F(13)F^+(34)G^{-1}(42)d(34). \tag{2.80}$$

In addition, we will need equations of motion for the anomalous Green's functions which would enable us to obtain an approximation to the one–particle Green's function.

In these two examples we made an assumption about the nature of the second–order propagator which enables us to obtain an approximation for the first–order propagator. In order to approximate the two–particle propagator one must decouple it from the third and higher order propagators by an appropriate ansatz. We now derive the *Bethe–Salpeter equation* and the *Brueckner approximation* which are two widely used methods of moving beyond the 1–propagator and the Hartree–Fock or Hartree–Fock–Bogolubov approximations.

The equation of motion for $G_2$, which involves $G_3$, is as follows:

$$\left[i\frac{\partial}{\partial t_1} - K(\nabla_1) + \mu(1)\right] G_2(1234) \mp i\int \Phi(15)G_3(125534)d(5) =$$

$$= \delta(14)G(23) \pm \delta(13)G(24). \tag{2.81}$$

Using (2.68) the first term in this equation can be rewritten as

$$\left[i\frac{\partial}{\partial t_1} - K(\nabla_1) + \mu(1)\right] G_2(1234) =$$

$$= \int \left[G^{-1}(15) + \Sigma(15)\right] G_2(5234)d(5).$$

Operating on (2.81) by $G$, leads to the integral equation

$$G_2(1234) = G(14)G(23) \pm G(13)G(24) + i\int G(15)\Phi(56)G_3(256634)d(56) -$$

$$- \int G(15)\Sigma(56)G_2(6234)d(56). \tag{2.82}$$

Unhappily this involves $G_3$, otherwise it would be an equation determining $G_2$. The standard procedure is to "decouple" by an *ansatz* for $G_3$ in terms of $G = G_1$ and $G_2$. In analogy with (2.75), we assume that

$$G_3(123456) = G(16)G_2(2345) \pm G(15)G_2(2346) + G(14)G_2(2356), \qquad (2.83)$$

which is an expression with appropriate symmetry properties. Substituting (2.83) in (2.82) leads to the integral equation

$$G_2(1234) = G_2^0(1234) + i \int G(15)\Phi(56)G(26)G_2(5634)d(56)\pm$$

$$\pm \int G(15)\Sigma(56) \left[ G_2^0(2634) - G_2(2634) \right] d(56). \qquad (2.84)$$

In this we used the definition (2.66) for the self–energy and the notation

$$G_2^0(1234) := G(14)G(23) \pm G(13)G(24). \qquad (2.85)$$

Since the difference between $G_2$ and $G_2^0$ arises from interaction between particles, the last term in (2.84) includes the interaction potential $\Phi$ to the second degree and higher. Dropping the third term on the right–hand side of (2.84) leads to the *Bethe–Salpeter equation*

$$G_2 (1234) = G_2^0 (1234) + i \int G(15)\Phi (56) G (26) G_2 (5634) d (56) , \qquad (2.86)$$

which includes interaction in first order.

One could try to solve this equation directly or adopt an approach due to Brueckner by introducing the so–called $T$–matrix which is defined implicitly by the relation

$$\Phi (12) G_2 (1234) = \int T (1256) G_2^0 (5634) d (56) . \qquad (2.87)$$

Then from (2.86) and (2.87), using the property

$$\int G_2^0 (1256) G^{-1} (53) G^{-1} (64) d (56) = \delta (14) \delta (23) \pm \delta (13) \delta (24) ,$$

it is possible to show that

$$T (1234) = \Phi (12) \delta (13) \delta (24) + i\Phi (12) \int G (15) G (26) T (5634) d (56) . \qquad (2.88)$$

The approach based on (2.88) is called the *Brueckner approximation*. If (2.88) can be solved for $T$, then (2.87) can be used to obtain $G_2$.

In concluding this section dealing with the equations of motion, we shall prove an important – even somewhat startling – result which echos the famous theorem of Kohn–Hohenberg. From (2.65) assuming that $t_1 \neq t_2$, we have

$$\int \Phi (13) G_2 (1332) d (3) = \mp \left[ i\frac{\partial}{\partial t_1} - K (\nabla_1) + \mu (1) \right] G (12) .$$

If, as in the standard case, the interaction betwen particles is symmetric so $V(x_1, x_2) = \Phi(x_1, x_2)$, then the average of the Hamiltonian (2.25) can be cast in the form

$$< H > = \pm \frac{i}{2} \int \lim_{(21)} \left[ i \frac{\partial}{\partial t_1} + K(\nabla_1) - \mu(1) \right] G(12)\, dx_1. \qquad (2.89)$$

Further, the internal energy (2.26) becomes

$$E = \pm \frac{i}{2} \int \lim_{(21)} \left[ i \frac{\partial}{\partial t_1} + K(\nabla_1) + U(x_1, t_1) + \mu \right] G(12)\, dx_1. \qquad (2.90)$$

For equilibrium, when $U(x, t) = U(x)$, we may invoke the Fourier transform in (2.90) to obtain

$$E = \pm \frac{i}{2} \int dx \int_{-\infty}^{+\infty} \left[ \omega + K(\nabla) + U(x) + \mu \right] e^{+i\omega 0} G(x, \omega) \frac{d\omega}{2\pi}, \qquad (2.91)$$

where the notation (2.52) is used. Finally, using the spectral representation (2.46) we obtain from (2.91)

$$E = \frac{1}{2} \int dx \int_{-\infty}^{+\infty} \left[ \omega + K(\nabla) + U(x) + \mu \right] J(x, \omega) n(\omega) \frac{d\omega}{2\pi}, \qquad (2.92)$$

where $J(x, \omega)$ is the diagonal element (2.57) of the spectral function (2.44).

This means that in order to calculate the internal energy of a system in equilibrium we need to know only the 1–propagator, or the diagonal of the corresponding spectral function. Note that the expressions (2.91) and (2.92) are exact. Of course, it will only be for moderately simple models that we shall be able to obtain the exact 1–propagator.

## 2.5  Wave Functions

In this section we shall introduce a family of 1–particle functions which can be thought of as Fourier transforms of time–dependent natural orbitals. Since, in quantum mechanics, time and energy are conjugate observables the *effective wave functions* which we shall define will be associated with 1–particle energies. The effective Hamiltonian of which these wave functions are eigen-functions contains the self-energy operator and therefore, in analogy with Hartree–Fock orbitals, they may be thought of as describing "dressed" quasi–particles which encapsulate a major portion of the correlation between the bare particles. As far as we are aware the precise relation between natural orbitals and the effective wave functions has never been studied so we do not assert that the effective wave functions in the limit specified in (2.9) become the natural orbitals of the Hartree–Fock approximation to the state of the system. We do not rule this out as a possibility but will be surprised if it proves to be the case since, while the Hartree–Fock approximation and the method of effective Hamiltonian of this section are both attempts to take into account a portion of the interaction energy, they are probably not strictly equivalent. We believe that this is an issue which deserves further study.

For a system in equilibrium we reformulate the equations of motion in terms of their Fourier transforms. Thus the inverse propagator can be expanded as

$$G^{-1}(12) = \int_{-\infty}^{+\infty} G^{-1}(x_1, x_2, \omega) e^{-i\omega t} \frac{d\omega}{2\pi} \,, \tag{2.93}$$

in which as in (2.39) we have used the notation $t = t_{12}$. Then the evolution equation (2.69) takes the form

$$\int G^{-1}\left(x_1, x_3, \omega\right) G\left(x_3, x_2, \omega\right) dx_3 = \delta\left(x_1 - x_2\right). \tag{2.94}$$

Similarly for the self–energy,

$$\Sigma\left(12\right) = \int_{-\infty}^{+\infty} \Sigma\left(x_1, x_2, \omega\right) \frac{d\omega}{2\pi} \,. \tag{2.95}$$

Introducing the notation

$$H\left(x, x', \omega\right) := [K\left(\nabla\right) + U\left(x\right)]\delta(x - x') + \Sigma(x, x', \omega), \tag{2.96}$$

for the Fourier transform of the inverse propagator (2.68) we find

$$G^{-1}(x, x', \omega) = (\omega + \mu)\delta(x - x') - H\left(x, x', \omega\right). \tag{2.97}$$

With respect to a complete orthonormal set of one–particle functions, $\{\vartheta_n(x)\}$, we have expansions

$$G(x, x', \omega) = \sum_{m,n} G_{mn}(\omega)\vartheta_m(x)\vartheta_n^*(x') \,,$$

$$G^{-1}(x, x', \omega) = \sum_{m,n} G_{mn}^{-1}(\omega)\vartheta_m(x)\vartheta_n^*(x') \,, \tag{2.98}$$

and hence, by (2.97),

$$G_{mn}^{-1}\left(\omega\right) = \left(\omega + \mu\right)\delta_{mn} - H_{mn}\left(\omega\right), \tag{2.99}$$

where

$$H_{mn}(\omega) := \int \vartheta_m^*(x) H\left(x, x', \omega\right)\vartheta_n(x')dx dx'. \tag{2.100}$$

The equation of motion (2.94) becomes

$$\sum_p G_{mp}^{-1}(\omega)G_{pn}(\omega) = \delta_{mn}. \tag{2.101}$$

Similarly, we can express the spectral function (2.49) in the form

$$J(x, x', \omega) = \sum_{m,n} J_{mn}(\omega)\vartheta_m(x)\vartheta_n^*(x'), \tag{2.102}$$

with

$$J_{mn}\left(\omega\right) = i\left[G_{mn}\left(\omega + i0\right) - G_{mn}\left(\omega - i0\right)\right].$$

In terms of these we may write the corresponding spectral representations and dispersion relations.

Expressions such as the above are most useful if $G_{mn}(\omega)$ is a diagonal matrix. We define an *effective Hamiltonian*, $H(x,\omega)$, as an integral operator such that

$$H(x,\omega)\varphi(x,\omega) := \int H(x,x',\omega)\varphi(x',\omega)\,dx', \tag{2.103}$$

with kernel (2.96). From (2.94) and (2.97) the equation of motion then takes the form

$$[\omega + \mu - H(x,\omega)]G(x,x',\omega) = \delta(x - x'). \tag{2.104}$$

If we substitute (2.98) in this and assume that $\vartheta_n$ can be chosen so that $G_{mn}(\omega) = G_m(\omega)\delta_{mn}$, we find

$$\sum_n G_n(\omega)[\omega + \mu - H(x,\omega)]\vartheta_n(x)\vartheta_n^*(x') = \delta(x - x') = \sum_n \vartheta_n(x)\vartheta_n^*(x').$$

In the last step we have used the assumption that $\{\vartheta_n\}$ is a complete orthonormal set. This equality will be possible only if

$$G_n(\omega)[\omega + \mu - H(x,\omega)]\vartheta_n(x) = \vartheta_n(x).$$

In particular it is clear that the $\vartheta_n$ are eigenfunctions of $H(x,\omega)$ with eigenvalues depending only on $\omega$ and the constant $\mu$. Set

$$H(x,\omega)\vartheta_n(x) = H_n(\omega)\vartheta_n(x). \tag{2.105}$$

Equation (2.105) is called the *effective wave equation* and the functions $\vartheta_n$, *effective wave functions*. Some authors also call (2.105) the *effective Schrödinger equation*. Comparing (2.104) and (2.105), we see that the poles of the propagator are defined by the eigenvalues of the effective Hamiltonian. The eigenfunctions, $\vartheta_n(x)$, describe quantum states, indexed by $n$, analogous to the familiar wave functions for the 1–particle Schrödinger equation. However, the effective Schrödinger equation is an *integral* rather than a differential equation and contains the additional parameter $\omega$.

Because of (2.105), the matrix elements (2.100) take the form

$$H_{mn}(\omega) = H_n(\omega)\delta_{mn}. \tag{2.106}$$

The Fourier coefficients of the inverse propagator (2.99) then become

$$G_{mn}^{-1}(\omega) = [\omega + \mu - H_n(\omega)]\delta_{mn}. \tag{2.107}$$

Substituting (2.107) into the equation of motion (2.101), we obtain

$$G_{mn}(\omega) = G_n(\omega)\delta_{mn}, \qquad G_n(\omega) = \frac{1}{\omega - H_n(\omega) + \mu}. \tag{2.108}$$

Equation (2.108) defines $G_n(\omega)$ as a function of $\omega$ on the whole of the complex plane apart from the real axis. For real values of $\omega$, $G_n$ is defined by means of the spectral representation (2.46). This, together with the spectral function $J_{mn}(\omega) = J_n(\omega)\delta_{mn}$, yields

$$G_n(\omega) = \int_{-\infty}^{+\infty} J_n(\omega) \left[ \frac{1 \pm n(\omega')}{\omega - \omega' + i0} \mp \frac{n(\omega')}{\omega - \omega' - i0} \right] \frac{d\omega'}{2\pi}. \qquad (2.109)$$

When the spectral function, $J_n(\omega)$, has no poles on the real axis, (2.109) implies that

$$J_n(\omega) = i[G_n(\omega + i0) - G_n(\omega - i0)]. \qquad (2.110)$$

Indeed, in the expression in the square bracket of (2.109), the terms involving $n(\omega')$ cancel and leave an integrand $J_n(\omega)/(\omega - \omega')$ to which Cauchy's formula applies.

In order to calculate $J_n(\omega)$, we need to know how (2.106) depends on $\omega$, so in view of (2.96) we separate the self–energy into two parts, which are respectively independent and dependent on $\omega$, by defining $\Lambda(x, x', \omega)$ to satisfy

$$\Sigma(x, x', \omega) = \Sigma(x, x', 0) + \Lambda(x, x', \omega). \qquad (2.111)$$

Then , in view of (2.103) and (2.105),

$$H_n(\omega) = H_n(0) + \Lambda_n(\omega), \qquad (2.112)$$

where

$$H_n(0) = \int \vartheta_n^*(x) \left[K(\nabla) + U(x)\right] \vartheta_n(x)dx + \int \vartheta_n^*(x)\Sigma(x, x', 0)\vartheta_n(x')dxdx',$$

and

$$\Lambda_n(\omega) = \int \vartheta_n^*(x)\Lambda(x, x', \omega)\vartheta_n(x')dxdx'.$$

The second term in (2.112) can be related to the propagator $G_n(\omega)$ by means of the equation of motion and has analytical properties similar to those of $G_n(\omega)$. Just as $J_n(\omega)$ was defined in connection with the spectral representation of $G_n(\omega)$, so we may introduce a spectral function, $\Gamma_n(\omega)$ such that for complex values of $\omega$, with $\mathrm{Im}(\omega) \neq 0$,

$$\Lambda_n(\omega) = \int_{-\infty}^{+\infty} \frac{\Gamma_n(\omega')}{\omega - \omega'} \frac{d\omega'}{2\pi}. \qquad (2.113)$$

For real $\omega$, the spectral representation is

$$\Lambda_n(\omega) = \int_{-\infty}^{+\infty} \Gamma_n(\omega') \left[ \frac{1 \pm n(\omega')}{\omega - \omega' + i0} \mp \frac{n(\omega')}{\omega - \omega' - i0} \right] \frac{d\omega'}{2\pi}. \qquad (2.114)$$

In analogy with (2.110), it can be proved that

$$\Gamma_n(\omega) = i\left[\Lambda_n(\omega + i0) - \Lambda_n(\omega - i0)\right]. \qquad (2.115)$$

Thus, when we take (2.112) and (2.113) into account, the propagator (2.108) assumes, for complex $\omega$ off the real axis, the value

$$G_n(\omega) = \left\{ \omega - H_n(0) + \mu - \int_{-\infty}^{+\infty} \frac{\Gamma_n(\omega')}{\omega - \omega'} \frac{d\omega'}{2\pi} \right\}^{-1}. \qquad (2.116)$$

In preparation for applications, we define an *effective energy* as

$$E_n(\omega) := H_n(0) + P \int_{-\infty}^{+\infty} \frac{\Gamma_n(\omega')}{\omega - \omega'} \frac{d\omega'}{2\pi}. \qquad (2.117)$$

Then substituting (2.116) in (2.110) we obtain the following important and very useful expression for the spectral function

$$J_n(\omega) = \frac{\Gamma_n(\omega)}{[\omega - E_n(\omega) + \mu]^2 + \Gamma_n^2(\omega)/4}. \qquad (2.118)$$

The expression

$$\mathrm{Re}\Lambda_n(\omega) = P \int_{-\infty}^{+\infty} \frac{\Gamma_n(\omega')}{\omega - \omega'} \frac{d\omega'}{2\pi} \qquad (2.119)$$

is called the *energy shift*, and

$$\Gamma_n(\omega) = -\frac{2\mathrm{Im}\Lambda_n(\omega)}{1 \pm 2n(\omega)} \qquad (2.120)$$

is called the *decay rate* for the state $\vartheta_n(x)$.

Neglecting the decay is equivalent to retaining only the real part of the poles of the propagator. In this case, (2.110) together with (2.47) imply that

$$J_n(\omega) = 2\pi \delta(\omega - E_n(\omega) + \mu). \qquad (2.121)$$

When a smooth function $f(\omega)$ has only simple zeros, $\omega_i$, that is $f'(\omega_i) \neq 0$ for all $i$,

$$\delta(f(\omega)) = \sum_i \frac{\delta(\omega - \omega_i)}{|f'(\omega_i)|}.$$

Applying this formula to (2.121) yields

$$J_n(\omega) = \sum_i \frac{2\pi \delta(\omega - \omega_{ni})}{|1 - E_n'(\omega_{ni})|}, \qquad (2.122)$$

where $\omega_{ni}$ are defined by the equation

$$\omega_{ni} = E_n(\omega_{ni}) - \mu, \qquad (i = 1, 2, \ldots)$$

and

$$E_n'(\omega) = \frac{d}{d\omega} E_n(\omega); \qquad E_n'(\omega_{ni}) \neq 1.$$

If we use the spectral function (2.122), the propagator (2.109) takes the form

$$G_n(\omega) = \sum_i \frac{1}{|1 - E_n'(\omega_{ni})|} \left[ \frac{1 \pm n(\omega_{ni})}{\omega - \omega_{ni} + i0} \mp \frac{n(\omega_{ni})}{\omega - \omega_{ni} - i0} \right], \qquad (2.123)$$

which is simpler and easier to use than when the spectral function (2.118) is inserted in (2.109). But one should remember that this last form is appropriate only if decay processes may validly be neglected.

## 2.6 Scattering Matrix

In his very first paper about Quantum Mechanics, Heisenberg presented a theory in terms of infinite matrices which, as far as possible, involved only observable properties of the physical situation under examination. Later, he introduced his *scattering matrix* appropriate for a situation in which one or more particles came into a "black box" and emerged in a different direction or as different species. He defined a matrix which connected the intial to the final state of the system. In this section, we use propagators to take into account the nature of the "black box" and to calculate the scattering matrix. Our $S_\infty$, defined in (2.139), is a particular case of Heisenberg's concept. The ideas of this section play an important role in providing a practical method of calculating Green's functions.

We envisage a system which is evolving under the control of a Hamiltonian $H$ and ask how it behaves if the Hamiltonian is modified by the addition of a term, $\Delta H$, which may or may not be "small". Both $H$ and $\Delta H$ are functionals of the field operators, which we indicate by

$$H = H\{\psi\}, \qquad \Delta H = \Delta H\{\psi\}. \qquad (2.124)$$

The new Hamiltonian is

$$H_\Delta := \tilde{H} + \Delta\tilde{H}, \qquad (2.125)$$

where

$$\tilde{H} = H\{\psi_\Delta\}, \qquad \Delta\tilde{H} = \Delta H\{\psi_\Delta\} \qquad (2.126)$$

are the same functionals as in (2.124) but functionals of new field operators $\psi_\Delta$ which satisfy the Heisenberg equation

$$i\frac{\partial}{\partial t}\psi_\Delta(x,t) = [\psi_\Delta(x,t), H_\Delta(t)]. \qquad (2.127)$$

Thus, the time dependence of the new field operators is given by

$$\psi_\Delta(x,t) = U_\Delta^+(t)\,\psi_\Delta(x,0)\,U_\Delta(t), \qquad (2.128)$$

where the evolution operator satisfies the equation

$$i\frac{d}{dt}U_\Delta(t) = H_\Delta U_\Delta(t),$$

with $H_\Delta$ and $H_\Delta(t)$ related, analogously to (2.128), by

$$H_\Delta(t) = U_\Delta^+(t) H_\Delta U(t).$$ (2.129)

Since the statistics of particles is independent of the Hamiltonian, the new field operators satisfy the same commutation relations as the old ones. In other words,

$$[\psi_\Delta(x,t), \psi_\Delta(x',t)]_\mp = 0,$$

and

$$\left[\psi_\Delta(x,t), \psi_\Delta^\dagger(x',t)\right]_\mp = \delta(x - x').$$

Since $\tilde{H}$ will be assumed to be self–adjoint and we want the creation operators $\psi_\Delta^\dagger(x,t)$ to satisfy the same equation (2.127) as $\psi_\Delta(x,t)$, we shall also assume that $\Delta\tilde{H}$ is self–adjoint. Thus

$$\Delta\tilde{H}^+ = \Delta\tilde{H}.$$ (2.130)

This imposes a restriction on $\Delta\tilde{H}$ so that if, for example,

$$\Delta\tilde{H} := -\int \psi_\Delta^\dagger(x)\Delta\mu(x,x',t)\psi_\Delta(x')dxdx',$$ (2.131)

in which $\psi_\Delta(x) := \psi_\Delta(x,0)$ and $\Delta\mu$ is a complex–valued function, then (2.130) implies that

$$\Delta\mu^*(x,x',t) = \Delta\mu(x,x',t).$$ (2.132)

Corresponding to the new field operators, new propagators can be defined just as the old ones were in (2.3), (2.11), and (2.13). Thus, for the one– and two–particle propagators we have, respectively

$$\tilde{G}(12) := -i < \hat{T}\psi_\Delta(1)\psi_\Delta^\dagger(2) >,$$

$$\tilde{G}_2(1234) := - < \hat{T}\psi_\Delta(1)\psi_\Delta(2)\psi_\Delta^\dagger(3)\psi_\Delta^\dagger(4) > .$$ (2.133)

The statistical averaging in (2.133) does not presuppose that the system is in equilbrium. It is taken with respect to the system density matrix, $\hat{\rho}(t)$, prescribed by $\hat{\rho}(0)$ at initial time zero.

As compared to the equations of motion for the old propagators, those for the new ones, (2.133), contain an additional term which results from the commutation of $\psi_\Delta$ with

$$\Delta\tilde{H}(t) = U_\Delta^+(t) \Delta\tilde{H} U_\Delta(t).$$

For the one–particle propagator $\tilde{G}$, the equation of motion takes the form

$$i\frac{\partial}{\partial t_1}\tilde{G}(12) = \delta(12) - i < \hat{T}[\psi_\Delta(1), H_\Delta(t_1)]\psi_\Delta^\dagger(2) > .$$ (2.134)

If we take the additive term in the form (2.131) we find

$$\left[\psi_\Delta(x,t), \Delta\tilde{H}(t)\right] = -\int \Delta\mu(x,x',t)\psi_\Delta(x',t)dx',$$

and obtain

$$\left[i\frac{\partial}{\partial t_1} - K\left(\nabla_1\right) + \mu\left(1\right)\right] \tilde{G}\left(12\right) + \int \Delta\mu\left(x_1, x_3, t_1\right)\delta\left(t_1 - t_3\right)\tilde{G}\left(32\right)d\left(3\right) \mp$$

$$\mp i\int \Phi\left(13\right)\tilde{G}_2\left(1332\right)d\left(3\right) = \delta\left(12\right). \tag{2.135}$$

Notice that when $\Delta\mu \to 0$, so $\Delta\tilde{H}$ tends to zero,

$$\psi_\Delta\left(x, t\right) \to \psi\left(x, t\right), \qquad H_\Delta \to \tilde{H} \to H,$$

while (2.135) becomes (2.65). Since we have not assumed that $\Delta\tilde{H}$ is small, the equation (2.135) for the propagator is valid for any $\Delta\tilde{H}$ of the form (2.131).

We now introduce the *scattering matrix*

$$S\left(t_1, t_2\right) := \hat{T}\exp\left\{-i\int_{t_2}^{t_1} \Delta H\left(t\right)dt\right\}, \tag{2.136}$$

in which

$$\Delta H\left(t\right) = U^+\left(t\right)\Delta H U\left(t\right),$$

and the term

$$\Delta H := -\int \psi^\dagger\left(x\right)\Delta\mu\left(x, x', t\right)\psi\left(x'\right)dx\,dx' \tag{2.137}$$

has the same form as $\Delta\tilde{H}$ in (2.131) but contains the original field operators $\psi$ and $\psi^\dagger$.

The scattering matrix, $S(t_1, t_2)$, has the following properties:

$$S(t, t) = 1, \qquad \lim_{\Delta H \to 0} S\left(t_1, t_2\right) = 1,$$

$$S(t_1, t_2)S\left(t_2, t_3\right) = S\left(t_1, t_3\right), \tag{2.138}$$

$$S^+\left(t_1, t_2\right) = S^{-1}\left(t_1, t_2\right) = S\left(t_2, t_1\right).$$

The limits of integration, $t_1$ and $t_2$ in (2.136), can be any real numbers. In particular we may define the balanced double limit

$$S_\infty := \lim_{t \to +\infty} S\left(t, -t\right) = S\left(+\infty, -\infty\right), \tag{2.139}$$

giving

$$S_\infty := \hat{T}\exp\left\{-i\int_{-\infty}^{+\infty} \Delta H\left(t\right)dt\right\}. \tag{2.140}$$

By considering the time–ordered product of $S_\infty$ with the field operator $\psi\left(1\right)$, we infer

$$\hat{T}S_\infty\psi\left(1\right) = S\left(+\infty, t_1\right)\psi\left(1\right)S\left(t_1, -\infty\right).$$

In a similar manner, we can obtain the time–ordered product of $S_\infty$ with several field operators. For example,

$$\hat{T}S_\infty\psi(1)\psi^\dagger(2) = \Theta(t_{12})S(+\infty, t_1)\psi(1)S(t_1, t_2)\psi^\dagger(2)S(t_2, -\infty)\pm$$

$$\pm \Theta(t_{21}) S(+\infty, t_2) \psi^\dagger(2) S(t_2, t_1) \psi(1) S(t_1, -\infty).$$

With the help of the scattering matrix we can express the new propagators in (2.133) by means of the old field operators:

$$\tilde{G}(12) = -i \frac{< \hat{T} S_\infty \psi(1) \psi^\dagger(2) >}{< S_\infty >},$$

$$\tilde{G}_2(1234) = -\frac{< \hat{T} S_\infty \psi(1) \psi(2) \psi^\dagger(3) \psi^\dagger(4) >}{< S_\infty >}. \tag{2.141}$$

Similarly, the propagator, $\tilde{G}_n$, of order $n$, can be expressed by means of the field operators as

$$\tilde{G}_n(1 \ldots 2n) = (-i)^n \frac{< \hat{T} S_\infty \psi(1) \ldots \psi(n) \psi^\dagger(n+1) \ldots \psi^\dagger(2n) >}{< S_\infty >}. \tag{2.142}$$

When $\psi$ in (2.13) is replaced by $\psi_\Delta$, we obtain (2.142). The proof is obtained by considering their equations of motion. We illustrate this in the case of the first order propagator in (2.141). Differentiate $\tilde{G}$ with respect to $t_1$ and take into account the relations

$$i \frac{\partial}{\partial t_1} S(+\infty, t_1) = -S(+\infty, t_1) \Delta H(t_1),$$

$$i \frac{\partial}{\partial t_1} S(t_1, t_2) = \Delta H(t_1) S(t_1, t_2),$$

$$i \frac{\partial}{\partial t_1} S(t_2, t_1) = -S(t_2, t_1) \Delta H(t_1),$$

$$i \frac{\partial}{\partial t_1} S(t_1, -\infty) = \Delta H(t_1) S(t_1, -\infty).$$

From (2.141) we then find

$$i \frac{\partial}{\partial t_1} \tilde{G}(12) = \delta(12) - i \frac{< \hat{T} S_\infty [\psi(1), H(t_1) + \Delta H(t_1)] \psi^\dagger(2) >}{< S_\infty >}.$$

This is seen to reduce to (2.135), when we recall the definition of $H(t)$ and note that (2.137) implies that

$$[\psi(x, t), \Delta H(t)] = -\int \Delta \mu(x, x', t) \psi(x', t) dx'.$$

It is important to realize that the representation (2.141) for the propagators (2.133) is valid for arbitary shifts $\Delta H$ of the Hamiltonian (2.125) whether or not the system is in equilibrium. It is frequently assumed that $H$ is the Hamiltonian for non–interacting particles while $\Delta H$ describes their interaction. In this case, (2.141) is referred to as the *interaction representation* for the Green's functions.

Consider a particular case for which $\Delta H$ is so small that we may restrict ourselves to terms which are linear in

$$\Delta G\,(12) := \tilde{G}\,(12) - G\,(12)\,. \qquad (2.143)$$

For such an approximation, linear in $\Delta H$, the scattering matrix (2.140) is

$$S_\infty \simeq 1 - i\hat{T} \int_{-\infty}^{+\infty} \Delta H\,(t)\,dt\,. \qquad (2.144)$$

From (2.141), it follows that

$$\tilde{G}\,(12) \simeq G\,(12) + iG\,(12) \int_{-\infty}^{+\infty} <\hat{T}\Delta H\,(t)> dt-$$

$$- \int_{-\infty}^{+\infty} <\hat{T}\Delta H\,(t)\,\psi\,(1)\,\psi^\dagger\,(2)> dt\,.$$

If $\Delta H$ has the form (2.137), then

$$<\hat{T}\Delta H\,(t)> = \mp \int \Delta\mu\,(x_3, x_4, t)\,\delta\,(t_3 - t)\,\delta\,(t_4 - t)\,G\,(43)\,d\,(34)\,,$$

$$<\hat{T}\Delta H\,(t)\,\psi\,(1)\,\psi^\dagger\,(2)> = \pm \int \Delta\mu\,(x_3, x_4, t)\,\delta\,(t_3 - t)\,\delta\,(t_4 - t)\,G_2\,(1432)\,d\,(34)\,.$$

Therefore the difference (2.143) becomes

$$\Delta G\,(12) = \pm \int [G\,(12)\,G\,(43) - G_2\,(1432)]\,\Delta\mu\,(x_3, x_4, t_3)\,\delta\,(t_3 - t_4)\,d\,(34)\,.$$

This last expression, with the notation

$$\Delta\mu\,(12) := \Delta\mu\,(x_1, x_2, t_1)\,\delta\,(t_1 - t_2)\,, \qquad (2.145)$$

takes the form

$$\Delta G\,(12) \simeq \pm \int [G\,(12)\,G\,(43) - G_2\,(1432)]\,\Delta\mu\,(34)\,d\,(34)\,. \qquad (2.146)$$

Define the variational derivative

$$\frac{\delta G\,(12)}{\delta\mu\,(34)} := \lim_{\Delta\mu\to 0} \frac{\Delta G\,(12)}{\Delta\mu\,(34)}\,. \qquad (2.147)$$

Then (2.146) yields the *variational representation*

$$G_2\,(1234) = G\,(14)\,G\,(23) \mp \frac{\delta G\,(14)}{\delta\mu\,(32)}\,, \qquad (2.148)$$

for the two–particle propagator.

If $\Delta\mu$ is diagonal in $x$, so that

$$\Delta\mu\left(x, x', t\right) = \Delta\mu\left(x, t\right)\delta\left(x - x'\right),$$

then (2.146) gives

$$\Delta G\left(12\right) \simeq \pm \int \left[G(12)G\left(33\right) - G_2\left(1332\right)\right]\Delta\mu\left(3\right)d\left(3\right), \qquad (2.149)$$

where

$$\Delta\mu\left(1\right) = \Delta\mu\left(x_1, t_1\right).$$

Defining the variational derivative

$$\frac{\delta G\left(12\right)}{\delta\mu\left(3\right)} := \lim_{\Delta\mu\to 0}\frac{\Delta G\left(12\right)}{\Delta\mu\left(3\right)}, \qquad (2.150)$$

we obtain the variational representation

$$G_2\left(1332\right) = G\left(12\right)G\left(33\right) \mp \frac{\delta G\left(12\right)}{\delta\mu\left(3\right)} \qquad (2.151)$$

for the two–particle propagator. Note that this has precisely the form which we need in the evolution equation (2.65).

## 2.7   Perturbation Theory

Using the definition of the scattering matrix it is straightforward to develop perturbation theory for Green's functions.

Suppose that the *unperturbed Hamiltonian* in (2.124) is $H \equiv H_0$ with

$$H_0 = \int \psi^\dagger\left(x\right)\left[K\left(\nabla\right) - \mu\left(x, t\right)\right]\psi\left(x\right)dx, \qquad (2.152)$$

where $\mu\left(x, t\right)$ may include the chemical potential, external fields and some self–consistent mean–fields. It might, for example, include the Hartree potential (2.78). We assume a perturbation, corresponding to a two particles interaction, in the form

$$\Delta H = \frac{1}{2}\int \psi^\dagger\left(x\right)\psi^\dagger\left(x'\right)V\left(x, x'\right)\psi\left(x'\right)\psi\left(x\right)dx dx'. \qquad (2.153)$$

Denote the *unperturbed Green's function* by $G_0$. Employing the commutator

$$\left[\psi\left(1\right), H_0\left(t_1\right)\right] = \left\{K\left(\nabla_1\right) - \mu\left(1\right)\right\}\psi\left(1\right),$$

we obtain the equation of motion for the *unperturbed* propagator, $G_0$:

$$\left[i\frac{\partial}{\partial t_1} - K\left(\nabla_1\right) + \mu\left(1\right)\right]G_0\left(12\right) = \delta\left(12\right). \qquad (2.154)$$

This equation can be solved by the method of expansion in wave functons, discussed in Section 2.4.

We proceed as follows in order to obtain perturbative corrections to $G_0$. Expand the scattering matrix (2.140) in the series

$$S_\infty = 1 + \sum_{k=1}^{\infty} \frac{(-i)^k}{k!} \int_{-\infty}^{+\infty} \hat{T} \prod_{p=1}^{k} \Delta H\left(t_p\right) dt_p. \tag{2.155}$$

Truncating this series at order $k$ and substituting it into the interaction representation (2.141) gives rise to the $k$-th order approximation, $G^{(k)}$, to the propagator $\tilde{G}$. This approximation includes powers of $\Delta H$ up to the $k$-th. Clearly, the zero-th order approximation is the unperturbed propagator

$$G^{(0)}\left(12\right) := G_0\left(12\right).$$

The $k$-th order approximation, $G_n^{(k}$, for the $n$-particle propagator, $\tilde{G}_n$, can be obtained in the same way.

Taking account of the equality

$$\int_{-\infty}^{+\infty} < \tilde{T} \Delta H\left(t\right) > dt = - \int G_2^{(0)}\left(1221\right) \Phi\left(12\right) d\left(12\right),$$

the first-order approximation for the one-particle propagator becomes

$$G^{(1)}\left(12\right) = G_0\left(12\right) +$$

$$+i \int \left[G_3^{(0)}\left(134432\right) - G_0\left(12\right) G_2^{(0)}\left(3443\right)\right] \Phi\left(34\right) d\left(34\right). \tag{2.156}$$

Here, the $n$-particle propagators, $G_n^{(0)}$, satisfy equations of motion involving the unperturbed Hamiltonian, $H_0$.

The equation of motion for $G_2^{(0)}$ is as follows:

$$\left[i\frac{\partial}{\partial t_1} - K\left(\nabla_1\right) + \mu\left(1\right)\right] G_2^{(0)}\left(1234\right) = \delta\left(14\right) G_0\left(23\right) \pm \delta\left(13\right) G_0\left(24\right). \tag{2.157}$$

Since the solution of a first-order differential equation is determined by the intial values, by comparing this equation with (2.154) we may conclude that

$$G_2^{(0)}\left(1234\right) = G_0\left(14\right) G_0\left(23\right) \pm G_0\left(13\right) G_0\left(24\right). \tag{2.158}$$

In a similar manner, from the equation of motion for $G_3^{(0)}$ together with (2.154) and (2.157), we conclude that

$$G_3^{(0)}\left(123456\right) = G_0\left(16\right) G_2^{(0)}\left(2345\right) \pm$$

$$\pm G_0\left(15\right) G_2^{(0)}\left(2346\right) + G_0\left(14\right) G_2^{(0)}\left(2356\right). \tag{2.159}$$

From (2.158) it follows that $G_3^{(0)}$ can be expressed as a sum of six terms involving only the first–order propagator $G_0$. We symbolize this by

$$G_3^{(0)} = \sum_P (\pm 1)^P G_0 G_0 G_0,$$

where $P$ runs over the group of permutations of $\{4, 5, 6\}$. By $(\pm 1)^P$ we mean $+1$ for bosons and, for fermions, the parity, $+1$ or $-1$, of the permutation $P$. Each $G_0$ must contain a variable from the first three and from the second three variables in $G_3^{(0)}$. The terms which occur can be obtained from $G_0(16)G_0(25)G_0(34)$, in which $\{1, 2, 3\}$ are in natural order and $\{4, 5, 6\}$ are in reversed order, by the six possible permutations of $\{4, 5, 6\}$.

Similarly, any $n$–particle unperturbed propagator can be presented as a sum

$$G_n^{(0)} = \sum_P (\pm 1)^P G_0 G_0 \ldots G_0. \tag{2.160}$$

The sum will contain $n!$ terms obtained by a rule which is the obvious generalization of the preceding example. In quantum field theory, this result is usually referred to as *Wick's theorem*.

Using (2.158) and (1.159) in (2.156) leads us to

$$G^{(1)}(12) = G_0(12) \pm i \int G_0(13)\,\Phi(34)\,G_2^{(0)}(3442)\,d(34), \tag{2.161}$$

for the first–order approximation to $G(12)$. The second–order approximation is given by

$$G^{(2)}(12) = G^{(1)}(12) + \frac{1}{2}\int_{-\infty}^{+\infty} < \hat{T}\left\{\left[G_0(12) + i\psi(1)\psi^\dagger(2)\right] \times\right.$$

$$\left. \times\, [\Delta H(t_3) - 2 < \Delta H(t_3) >]\,\Delta H(t_4)\right\} > dt_3 dt_4,$$

in which we need to substitute (2.153) and make use of (2.160).

In a similar manner we can obtain any $k$–th order approximation for an $n$-particle propagator $G_n^{(k)}$.

Another approach to perturbation theory uses the Dyson equation (2.72). Setting

$$\Sigma_0(12) = 0, \tag{2.162}$$

we find that the inverse propagator (2.74) is

$$G_0^{-1}(12) = \left[i\frac{\partial}{\partial t_1} - K(\nabla_1) + \mu(1)\right]\delta(12), \tag{2.163}$$

and the kernel (2.73) becomes

$$B(12) = \int G_0(13)\,\Sigma(32)\,d(3).$$

The Dyson equation, (2.72), takes the form

$$G\left(12\right) = G_0\left(12\right) + \int G_0\left(13\right) \Sigma\left(34\right) G\left(42\right) d\left(34\right). \qquad (2.164)$$

Successive approximations to $G$ are obtained by iterating (2.164) according to the scheme

$$G^{(k)} \to \Sigma^{(k+1)} \to G^{(k+1)}. \qquad (2.165)$$

The first–order approximation gives

$$G^{(1)}\left(12\right) = G_0\left(12\right) + \int G_0\left(13\right) \Sigma^{(1)}\left(34\right) G_0\left(42\right) d\left(34\right), \qquad (2.166)$$

in which

$$\Sigma^{(1)}\left(12\right) = \pm i \int \Phi\left(13\right) G_2^{(0)}\left(1334\right) G_0^{-1}\left(42\right) d\left(34\right).$$

With $G_2^{(0)}$ given by (2.158), we have

$$\Sigma^{(1)}\left(12\right) = i\Phi\left(12\right) G_0\left(12\right) \pm i\delta\left(12\right) \int \Phi\left(13\right) G_0\left(33\right) d\left(3\right). \qquad (2.167)$$

Therefore (2.166) coincides with (2.161).

Note that (2.167) has the same apparent structure as the Hartree–Fock self–energy (2.76) but is quite different since $\Sigma^{(1)}$ involves $G_0$, while $\Sigma_{HF}$ involves $G$.

By (2.66) the Dyson equation (2.164) can be transformed into

$$G\left(12\right) = G_0\left(12\right) \pm i \int G_0\left(13\right) \Phi\left(34\right) G_2\left(3442\right) d\left(34\right). \qquad (2.168)$$

There is an equivalent approach to obtaining successive approximations to $G$ in which we adopt the recursive scheme

$$G^{(k)} \to G_2^{(k)} \to G^{(k+1)}. \qquad (2.169)$$

Substituting $G_2^{(0)}$ from (2.158) in the right–hand side of equation (2.168), again gives us (2.161). In order to find $G^{(2)}$, we need $G_2^{(1)}$. This can be obtained from either (2.141) or (2.65), giving

$$G_2^{(1)}\left(1234\right) = G_2^{(0)}\left(1234\right) +$$

$$+i \int_{-\infty}^{+\infty} < \hat{T}\left[G_2^{(0)}\left(1234\right) + \psi\left(1\right)\psi\left(2\right)\psi^{\dagger}\left(3\right)\psi^{\dagger}\left(4\right)\right] \Delta H\left(t\right) > dt.$$

With the help of (2.153), this yields

$$G_2^{(1)}\left(1234\right) = G_2^{(0)}\left(1234\right) +$$

$$+i \int \left[G_4^{(0)}\left(12566534\right) - G_2^{(0)}\left(1234\right) G_2^{(0)}\left(5665\right)\right] \Phi\left(56\right) d\left(56\right). \qquad (2.170)$$

The equation of motion for $G_4^{(0)}$, together with the Wick's theorem (2.160), yields

$$G_4^{(0)}\,(12345678) = G_0\,(18)\,G_3^{(0)}\,(234567) \pm G_0\,(17)\,G_3^{(0)}\,(234568) +$$

$$+G_0\,(16)\,G_3^{(0)}\,(234578) \pm G_0\,(15)\,G_3^{(0)}\,(234678)\,.$$

Invoking (2.159), we obtain

$$G_4^{(0)}\,(12345678) =$$

$$= G_2^{(0)}\,(1278)\,G_2^{(0)}\,(3456) \pm G_2^{(0)}\,(1378)\,G_2^{(0)}\,(2456) +$$

$$+G_2^{(0)}\,(1478)\,G_2^{(0)}\,(2356) + G_2^{(0)}\,(1256)\,G_2^{(0)}\,(3478) \pm$$

$$\pm G_2^{(0)}\,(1356)\,G_2^{(0)}\,(2478) + G_2^{(0)}\,(1456)\,G_2^{(0)}\,(2378)\,,$$

where $G_2^{(0)}$ is given by (2.158).

The two approaches to perturbation theory, are equivalent to one another, and applicable whether or not the system is in equilibrium, as long as $\Delta H$ can be considered small. Since Green's functions may be interpreted as generalized functions, convergence of the perturbation sequence is to be understood as convergence of functionals, – for example, as the convergence of the expected values for some local observables.

It may happen that terms in the perturbation expansion diverge when they involve integrals of products of several propagators. We therefore illustrate, by means of an example, two methods, which we call *implicit* and *explicit*, of avoiding such divergences [59].

Consider the Fourier transform of (2.163) for a system in equilibrium,

$$G_0^{-1}\,(x, x', \omega) = [\omega + \mu - H\,(x)]\,\delta\,(x - x')\,, \tag{2.171}$$

where

$$H\,(x) = K\,(\nabla) + U\,(x) \tag{2.172}$$

is hermitian. The propagator corresponding to (2.171) can be expanded in the form

$$G_0\,(x, x', \omega) = \sum_n G_n(\omega)\psi_n(x)\psi_n^*(x') \tag{2.173}$$

with respect to one–particle wave functions satisfying the equation

$$H\,(x)\,\psi_n\,(x) = E_n\psi_n\,(x)\,, \tag{2.174}$$

with expansion coefficients

$$G_n\,(\omega) = \frac{1}{\omega - E_n + \mu} \tag{2.175}$$

defined on the complex $\omega$–plane. As was explained in Section 2.4, on the real $\omega$–axis

$$G_n\,(\omega) = \frac{1 \pm n\,(E_n)}{\omega - E_n + i0} \mp \frac{n\,(E_n)}{\omega - E_n - i0}\,. \tag{2.176}$$

The eigenvalues $E_n$ are real since we assumed that $H(x)$ is hermitian.

Starting from the zero order approximation (2.173) for the propagator, in successive approximations we meet a variety of integrals with integrands containing products of factors such as (2.176). For example:

$$I_{mn} = \pm i \int_{-\infty}^{+\infty} e^{+i\omega 0} G_m(\omega) G_n(\omega) \frac{d\omega}{2\pi},$$

$$I'_{mn} = \pm i \int_{-\infty}^{+\infty} e^{+i\omega 0} \omega G_m(\omega) G_n(\omega) \frac{d\omega}{2\pi},$$

$$I_{m\ell n} = \pm i \int_{-\infty}^{+\infty} e^{+i\omega 0} G_m(\omega) G_\ell(\omega) G_n(\omega) \frac{d\omega}{2\pi},$$

$$I'_{m\ell n} = \pm i \int_{-\infty}^{+\infty} e^{+i\omega 0} \omega G_m(\omega) G_\ell(\omega) G_n(\omega) \frac{d\omega}{2\pi}. \tag{2.177}$$

The integrals in (2.177) are symmetric with respect to permutation of their indices.

When two or more of the indices in (2.177) coincide, a direct integration of $G_n^2(\omega)$, or in general of $G_n^k(\omega)$, would lead to a divergence. This would be due to the fact that the propagator $G_n(\omega)$ must be treated as a *generalized function* or *distribution* and products of distributions with coincident singualarities are not uniquely determined. They require additional definition.

There are two approaches, so–called *implicit* and *explicit*, which avoid the divergences [59].

($i$) For the *implicit* approach, instead of defining $G_n^k(\omega)$ directly we replace it by the limit of a sequence. We first calculate the integrals in (2.177) for distinct indices, giving

$$I_{mn} = A_{mn} + A_{nm},$$

$$I'_{mn} = E_m A_{mn} + E_n A_{nm},$$

$$I_{m\ell n} = B_{m\ell n} + B_{\ell nm} + B_{nm\ell},$$

$$I'_{m\ell n} = E_m B_{m\ell n} + E_\ell B_{\ell nm} + E_n B_{nm\ell},$$

where

$$A_{mn} = \frac{n(E_m)}{E_m - E_n},$$

and

$$B_{m\ell n} = \frac{n(E_m)[1 \pm n(E_\ell)][1 \pm n(E_n)] \pm [1 \pm n(E_m)]n(E_\ell)n(E_n)}{(E_m - E_\ell)(E_m - E_n)}.$$

We then define the integrals (2.177), for coincident indices, by the following limiting procedure:

$$I_{nn} := \lim_{m \to n} I_{mn} = -\beta n(E_n)[1 \pm n(E_n)],$$

$$I'_{nn} := \lim_{m \to n} I'_{mn} = n(E_n) + E_n I_{nn},$$

$$I_{n\ell n} := \lim_{m \to n} I_{m\ell n} = \frac{I_{nn} - I_{n\ell}}{E_n - E_\ell}, \tag{2.178}$$

$$I'_{n\ell n} := \lim_{m \to n} I'_{m\ell n} = \frac{E_n I_{nn} - E_\ell I_{n\ell}}{E_n - E_\ell},$$

$$I_{nnn} := \lim_{\ell \to n} I_{n\ell n} = \frac{\beta^2}{2} n\,(E_n)\,[1 \pm n\,(E_n)]\,[1 \pm 2n\,(E_n)],$$

$$I'_{nnn} := \lim_{\ell \to n} I'_{n\ell n} = I_{nn} + E_n I_{nnn}.$$

That is, we let the indices coincide only after integrating but not in the integrands.

($ii$) For the *explicit* approach, we first define the $k$-th power of $G_n$ as

$$G_n^k\,(\omega) := \frac{1 \pm n\,(\omega)}{(\omega - E_n + i0)^k} \mp \frac{n\,(\omega)}{(\omega - E_n - i0)^k}. \tag{2.179}$$

We then substitute the expression (2.179) directly into the integrands. If under the integral, it is multiplied by a function, $f(\omega)$ say, which is differentiable of order $k$ or more and vanishes sufficiently rapidly at infinity, then integration by parts gives

$$\pm i \int_{-\infty}^{+\infty} e^{+i\omega 0} f\,(\omega)\, G_n^k\,(\omega)\, \frac{d\omega}{2\pi} = \frac{1}{(k-1)!} \frac{d^{k-1}}{dE_n^{k-1}}\, [f\,(E_n)\, n\,(E_n)].$$

Using the relations

$$\frac{d}{d\omega} n\,(\omega) = -\beta n\,(\omega)\,[1 \pm n\,(\omega)],$$

and

$$\frac{d^2}{d\omega^2} n\,(\omega) = -\beta\,[1 \pm 2n\,(\omega)]\, \frac{d}{d\omega} n\,(\omega),$$

it is straightforward to show that these two approaches to defining the integrals (2.177) with coincident indices are equivalent.

Thus, by using either perturbation method, we obtain a sequence, $\{G^{(k)}\}$, of approximations for a propagator. With these we may then find a sequence, $\{< \hat{A} >_k\}$, of expected values for an element, $\hat{A}$, of an algebra of observables. For any particular problem, it will be the convergence properties of the sequences $< \hat{A} >_k$ that determine convergence of our procedure.

## 2.8 Excited States

It frequently happens that we are more interested in the spectrum of the quantum states of an equilibrium system than in its statistical states. As we proved in Section 2.4, the spectrum is given by the poles of the propagators in the $\omega$-representation and can be found from the effective wave equation (2.105).

The *effective Hamiltonian*

$$H\,(x, \omega) = H\,(x) + M\,(x, \omega) \tag{2.180}$$

consists of two terms. The first is the Schrödinger operator (2.172) and the second, called the *mass operator*, is an integral operator the kernel of which is the self–energy, $\Sigma$:

$$M\left(x,\omega\right)\varphi\left(x,\omega\right) := \int \Sigma\left(x,x',\omega\right)\varphi\left(x',\omega\right)dx'. \qquad (2.181)$$

Thus the effective wave equation is

$$\left[H\left(x\right) + M\left(x,\omega\right)\right]\varphi_n\left(x,\omega\right) = H_n\left(\omega\right)\varphi_n\left(x,\omega\right). \qquad (2.182)$$

This equation differs from the usual Schrödinger equation in three respects:

1. It is an integro–differential equation;

2. Since, in general, the effective Hamiltonian, (2.180), is not hermitian, the eigenvalues, $H_n(\omega)$, are not necessarily real;

3. The self–energy is normally defined only by means of the perturbation theory described in the previous section.

Therefore in order to solve (2.182) it will be necessary to use perturbation theory, but (3) implies that the familiar method of solving the Schrödinger equation is not applicable since this assumes that the Hamiltonian is known *before* recursion for successive approximations commences. We therefore attempt a solution by setting

$$M\left(x,\omega\right) = \sum_{k=0}^{\infty} \varepsilon^k M^{(k)}\left(x,\omega\right) \qquad (2.183)$$

in which $\varepsilon$ is a formal expansion parameter introduced to keep track of the successive orders of approximation. At the end of the calculation we let $\varepsilon \to 1$, hoping that the series converges! Each $M^{(k)}$ is an integral operator such that

$$M^{(k)}\left(x,\omega\right)\varphi\left(x,\omega\right) := \int M^{(k)}\left(x,x',\omega\right)\varphi\left(x',\omega\right)dx',$$

with the kernel, $M^{(k)}(x,x',\omega)$, defined by the corresponding approximation for the self–energy. In the zero approximation, $\Sigma_0(12) = 0$, we have

$$M^{(0)}\left(x,x',\omega\right) = \Sigma_0\left(x,x',\omega\right) = 0. \qquad (2.184)$$

The first approximation $M^{(1)}$ is given by

$$M^{(1)}\left(x_1,x_2,\omega\right) = \int_{-\infty}^{+\infty} \Sigma^{(1)}\left(12\right) e^{i\omega t} dt, \qquad (2.185)$$

with self–energy (2.167) and $t \equiv t_{12}$. Employing the expansion (2.173) for the zero–order propagator $G_0$, we find

$$i \int_{-\infty}^{+\infty} e^{+i\omega 0} G_0\left(x,x',\omega\right) \frac{d\omega}{2\pi} = \pm \rho_1^{(0)}\left(x,x'\right),$$

where

$$\rho_1^{(0)}\left(x, x'\right) := \sum_m n\left(E_m\right) \psi_m\left(x\right) \psi_m^*\left(x'\right) \tag{2.186}$$

is the 1–matrix in zero approximation and $E_k$ and $\psi_k$ are defined by (2.174). Thus (2.185) reduces to the expression

$$M^{(1)}(x, x', \omega) = \pm\Phi(x, x')\rho_1^{(0)}(x, x') + \delta(x - x') \int \Phi(x, x'')\rho^{(0)}(x'')dx'', \tag{2.187}$$

in which

$$\rho^{(0)}\left(x\right) := \rho_1^{(0)}\left(x, x\right) = \sum_m n\left(E_m\right) \left|\psi_m\left(x\right)\right|^2 \tag{2.188}$$

is the particle density. Note that (2.187) does not depend on $\omega$.

However, in higher approximations, that is for $k \geq 2$, the kernel

$$M^{(k)}\left(x_1, x_2, \omega\right) = \int_{-\infty}^{+\infty} \left[\Sigma^{(k)}\left(12\right) - \Sigma^{(k-1)}\left(12\right)\right] e^{i\omega t}dt \tag{2.189}$$

will, in general, depend on $\omega$.

The series (2.183) for the mass operator suggests that we should seek a solution of the eigenvalue problem (2.182) in the form

$$H_n\left(\omega\right) = E_n + \sum_{k=1}^{\infty} \varepsilon^k H_n^{(k)}\left(\omega\right),$$

$$\varphi_n\left(x, \omega\right) = \psi_n\left(x\right) + \sum_{k=1}^{\infty} \varepsilon^k \varphi_n^{(k)}\left(x, \omega\right). \tag{2.190}$$

The functions $\varphi_n^{(k)}$ can be expanded in the basis $\{\psi_n\}$, so that

$$\varphi_n^{(k)}\left(x, \omega\right) = \sum_m c_{mn}^{(k)}\left(\omega\right) \psi_m\left(x\right), \tag{2.191}$$

and, since we assume that the $\psi_m$ are orthonormal,

$$c_{mn}^{(k)}\left(\omega\right) = \left(\psi_m, \varphi_n^{(k)}\right), \qquad k \geq 1. \tag{2.192}$$

There is an additional condition on these coefficients expressing the fact that we want $\varphi_n(x, \omega)$ in (2.190) to be normalized,

$$\left(\varphi_n, \varphi_n\right) = 1. \tag{2.193}$$

Substitute the expansions (2.183) and (2.190) into (2.182) and equate like powers of $\varepsilon$. In zero order, this gives (2.174) and in first order:

$$[H(x) - E_n]\varphi_n^{(1)}(x, \omega) = \left[H_n^{(1)}(\omega) - M^{(1)}(x, \omega)\right]\psi_n(x). \tag{2.194}$$

This can be considered as a nonhomogeneous equation defining $\varphi_n^{(1)}$. Using (2.191) we may transform (2.194) to

$$c_{mn}^{(1)}(\omega)(E_m - E_n) = \delta_{mn} H_n^{(1)}(\omega) - M_{mn}^{(1)}(\omega),\tag{2.195}$$

in which

$$M_{mn}^{(k)}(\omega) := \int \psi_m^*(x) M^{(k)}(x, x', \omega)\psi_n(x')dxdx'.\tag{2.196}$$

The coefficients $c_{nn}^{(1)}(\omega)$, when $m = n$, are not specified by (2.195) but can be obtained from the normalization condition (2.193). If we substitute in this condition the expansion (2.191) and equate equal powers of $\varepsilon$, we find

$$c_{nn}^{(1)}(\omega) = \left(\psi_n, \varphi_n^{(1)}\right) = 0.\tag{2.197}$$

It follows that the first order corrections to the eigenvalues and eigenfunctions are

$$H_n^{(1)}(\omega) = M_{nn}^{(1)}(\omega),$$

$$\varphi_n^{(1)}(\omega) = \sum_m c_{mn}^{(1)}(\omega)\psi_m(x),\tag{2.198}$$

where

$$c_{mn}^{(1)}(\omega) = -(1 - \delta_{mn})\frac{M_{mn}^{(1)}(\omega)}{E_m - E_n}.\tag{2.199}$$

In the second order, we find

$$[H(x) - E_n]\varphi_n^{(2)}(x, \omega) = \left[H_n^{(1)}(\omega) - M^{(1)}(x, \omega)\right]\varphi_n^{(1)}(x, \omega) +$$

$$+ \left[H_n^{(2)}(\omega) - M^{(2)}(x, \omega)\right]\psi_n(x).\tag{2.200}$$

Together with (2.191) this leads to

$$c_{mn}^{(2)}(\omega)(E_m - E_n) = \delta_{mn} H_n^{(2)}(\omega) + c_{mn}^{(1)}(\omega) H_n^{(1)}(\omega) -$$

$$- \sum_\ell M_{m\ell}^{(1)}(\omega) c_{\ell n}^{(1)}(\omega) - M_{mn}^{(2)}(\omega).\tag{2.201}$$

From this we obtain

$$H_n^{(2)}(\omega) = -\sum_{m \neq n}\frac{M_{mn}^{(1)}(\omega) M_{nm}^{(1)}(\omega)}{E_m - E_n} + M_{nn}^{(2)}(\omega),$$

$$c_{mn}^{(2)}(\omega) = \sum_{\ell \neq n}\frac{M_{m\ell}^{(1)}(\omega) M_{\ell n}^{(1)}(\omega)}{(E_m - E_n)(E_\ell - E_n)} - \frac{M_{mn}^{(1)}(\omega) M_{nn}^{(1)}(\omega)}{(E_m - E_n)^2} - \frac{M_{mn}^{(2)}(\omega)}{E_m - E_n},\tag{2.202}$$

where in this last equation $m \neq n$. As before, $c_{nn}^{(2)}(\omega)$, with coincident indices, cannot be obtained from (2.201), but can be deduced from the normalization requirement (2.193) which, to within a phase factor, implies

$$c_{nn}^{(2)}(\omega) = -\frac{1}{2}\sum_{m \neq n}\left|\frac{M_{mn}^{(1)}(\omega)}{E_m - E_n}\right|^2.\tag{2.203}$$

Another method of defining the second-order correction, $H_n^{(2)}(\omega)$, is to multiply (2.200) by $\psi_n^*(x)$ and integrate the resulting equation with respect to $x$. Noting that $H(x)$ is hermitian and using (2.107) leads to

$$H_n^{(2)}(\omega) = \left(\psi_n, M^{(1)}\varphi_n^{(1)}\right) + M_{nn}^{(2)}(\omega),\tag{2.204}$$

where $\varphi_n^{(1)}$ is defined as the solution of the nonhomogeneous equation (2.194). Thus the correction, $H_n^{(2)}$, is expressed by means of quantities with the common label $n$. At first sight, (2.204) looks simpler than (2.202) which involves a summation. However, this second approach requires the solution of the differential equation (2.194) in order to evaluate the scalar product on the right hand side of (2.204). Which is the optimal approach will depend on the particular problem under consideration.

The procedures sketched above may be continued to any desired order of approximation. The approach based on the summation formulas, as in (2.198) and (2.202), is a version, in this context, of the standard *Rayleigh–Schrödinger perturbation theory*. The second approach, using (2.204) and solving a differential equation is analogous to the *Dalgarno-Lewis perturbation theory* [60,61] (see also [62,63]).

However as already mentioned, there are a number of differences between the problem of calculating the spectrum of a one–particle operator and that of obtaining the spectrum of excited states for a many–particle system. The most obvious is that the effective wave equation (2.182) contains the mass operator which is not given *a priori* but is itself defined recursively as the successive terms in (2.183) are obtained. Each $M^{(k)}$ is expressed in terms of the propagators, $G_k^{(0)}$ or the reduced density matrices, $\rho_k^{(0)}$. For example, $M^{(1)}$ contains the density matrix $\rho_1^{(0)}$ of (2.186), while $M^{(2)}$ contains the density matrix

$$\rho_2^{(0)}(x_1, x_2, x_3, x_4) = \mp \lim_{t_i \to t} G_2^{(0)}(1234),\tag{2.205}$$

where the limit is taken such that

$$t_4 > t_3 > t_1 > t_2.$$

For the one–particle propagator, with Fourier transform (2.98), we find

$$G_0(12) = -i\sum_n \{\Theta(t_{12})[1 \pm n(E_n)] \pm \Theta(-t_{12})n(E_n\} \times$$

$$\times \psi_n(x_1)\psi_n^*(x_2)\exp(-iE_n t_{12}).\tag{2.206}$$

Substituting (2.206) into (2.158) we obtain the two–particle propagator $G_2^{(0)}$. In view of (2.205), this defines the 2–matrix

$$\rho_2^{(0)}(x_1, x_2, x_3, x_4) = \sum_{m,n} n(E_m)n(E_n)\psi_m(x_1)\psi_n(x_2) \times$$

$$\times [\psi_n^*(x_3)\psi_m^*(x_4) \pm \psi_m^*(x_3)\psi_n^*(x_4)].\tag{2.207}$$

Defining the *two–particle wave function*

$$\psi_{mn}\left(x_1, x_2\right) := \frac{1}{\sqrt{2}}\left[\psi_m\left(x_1\right)\psi_n\left(x_2\right) \pm \psi_n\left(x_1\right)\psi_m\left(x_2\right)\right],$$

and the *pairon density of states*

$$B_{mn} := n\left(E_m\right)n\left(E_n\right),$$

the zero–order 2–matrix takes the form

$$\rho_2^{(0)}\left(x_1, x_2, x_3, x_4\right) = \sum_{m,n} B_{mn}\psi_{mn}\left(x_1, x_2\right)\psi_{nm}^{*}\left(x_3, x_4\right). \tag{2.208}$$

Another difference between the two eigenvalue problems arises from the fact the Hamiltonian (2.180) is not necessarily hermitian. Not only can this give rise to complex eigenvalues signalling the presence of decay, but it also precludes the use of the virial or hypervirial theorems which are widely employed in quantum mechanics. These theorems may be used in situations, discussed in Section 2.4, in which decay is negligible and the effective Hamiltonian can be approximated sufficiently accurately by a hermitian operator $H = K + U$.

If $K = -p^2/2m$ and $\mathbf{p} = i\nabla$, then if $H = K + U$ is hermitian with eigenfucntions $\psi_n$, then the *virial theorem* asserts that

$$2\left(\psi_n, K\psi_n\right) = \left(\psi_n, \left(\mathbf{r}\cdot\nabla U\right)\psi_n\right).$$

The proof of this is based on

$$[\mathbf{r} \cdot \mathbf{p}, H] = 2iK - i\mathbf{r}\cdot\nabla U,$$

and on the equality

$$\left(\psi_n, [\mathbf{r} \cdot \mathbf{p}, H]\psi_n\right) = 0.$$

The *hypervirial theorem* is a generalization of the virial theorem: If $H$ is a hermitian operator on Hilbert space with eigenfunctions, $\psi_n$, and if $A$ is a well–defined linear operator on the same space, then $\left(\psi_n, [A, H]\psi_n\right) = 0$. The proof uses the fact that $H$ is hermitian.

These theorems can sometimes be used to greatly simplify calculations in perturbation theory for the propagators but only if the effective Hamiltonian is strictly, or approximately, hermitian.

Perhaps the most important peculiarity of the many–body problem is that the eigenvalues of the effective equation (2.182) do not give us the spectrum of the excited states directly. This spectrum is obtained from the poles of the propagator (2.108) by solving the equation

$$\omega_n = H_n\left(\omega_n\right) - \mu \tag{2.209}$$

for $\omega_n$. This defines the *single–particle spectrum*. Similarly, one can find the poles of the two–particle propagator which define the *spectrum of the collective excitations*.

Expanding the two–particle propagator, in analogy with (2.98), one can derive a two–particle effective wave equation. One may then proceed to obtain $n$–particle effective wave equations of successive orders $n = 1, 2, 3, \ldots$, and attempt to solve them by a procedure similar to that described for (2.198). Note that the spectra for the effective Hamiltonians may depend on thermodynamic variables such as temperature.

## 2.9 Screened Potential

We remarked in Section 2.7, that the unperturbed Hamiltonian (2.152) may include potential fields. It is a widely used trick to include in such potentials an averaged form of the interactions among the particles. In this way, the convergence properties of the perturbation procedure can often be significantly improved.

An obvious candidate for a potential to be included in the unperturbed Hamiltonian for this purpose is the Hartree potential (2.78) which has the form

$$V_H (1) = \int \Phi (12) \, \rho (2) \, d (2) = \pm i \int \Phi (12) \, G (22) \, d (2) , \qquad (2.210)$$

where $\Phi(12)$, introduced in (2.63), is often referred to as the bare interaction potential. The unperturbed inverse propagator can then be written as

$$G_0^{-1} (12) = \left[ i \frac{\partial}{\partial t_1} - K (\nabla_1) - V (1) \right] \delta (12) , \qquad (2.211)$$

in which we have introduced the *effective potential*

$$V (1) := V_H (1) - \mu (1) . \qquad (2.212)$$

Thus, already in the zero approximation we partially take account of interaction among the particles by means of the self–consistent field (2.210). It is therefore appropriate to call this the *Hartree* or the *mean–field* approximation.

The total inverse propagator then takes the form

$$G^{-1} (12) = G_0^{-1}(12) - \Sigma' (12) , \qquad (2.213)$$

in which the supplemental term,

$$\Sigma' (12) := \Sigma (12) - V_H (1) \, \delta (12) , \qquad (2.214)$$

is called the *exchange–correlation self–energy* since it takes into account exchange and correlation effects not included in the Hartree approximation. We may then proceed further by following the perturbation scheme outlined in Section 2.7.

However, there is an alternative procedure which makes use of the variational derivatives discussed in Section 2.6. In following this route an important role is played by the *screened potential*

$$W (12) := \int \varepsilon^{-1} (13) \, \Phi (32) \, d (3) , \qquad (2.215)$$

in which

$$\varepsilon^{-1}(12) = -\frac{\delta V(1)}{\delta\mu(2)} \qquad (2.216)$$

is the so–called *inverse dielectric function*. The screened potential (2.215) is sometimes also called the *shielded potential*, and the function $\varepsilon^{-1}(12)$ in (2.216), the *dielectric response function*. Further, the solution, $\varepsilon(12)$, when it exists, of the integral equation

$$\int \varepsilon^{-1}(13)\,\varepsilon(32)\,d(3) = \delta(12),$$

is called the *dielectric function*. This is an appropriate name since it is a generalization of the familiar dielectric constant of electro–magnetic theory.

To obtain the inverse dielectric function we evaluate

$$\frac{\delta V(1)}{\delta\mu(2)} = -\delta(12) \pm i \int \Phi(13)\frac{\delta G(33)}{\delta\mu(2)}d(3),$$

and

$$\frac{\delta G(12)}{\delta\mu(3)} = \int \frac{\delta G(12)}{\delta V(4)}\frac{\delta V(4)}{\delta\mu(3)}d(4).$$

Introducing the *polarization function*

$$\Pi(12) := \frac{\delta\rho(1)}{\delta V(2)} = \pm i\frac{\delta G(11)}{\delta V(2)}, \qquad (2.217)$$

we finally have an integral equation

$$\varepsilon^{-1}(12) = \delta(12) + \int \Phi(13)\,\Pi(34)\,\varepsilon^{-1}(42)\,d(34) \qquad (2.218)$$

satisfied by the inverse dielectric function.

This equation must be supplemented by an equation for the polarization function (2.217). Varying the equation of motion (2.69) with respect to the effective potential (2.212), we are led to

$$\int \frac{\delta G^{-1}(13)}{\delta V(4)}G(32)d(3) + \int G^{-1}(13)\frac{\delta G(32)}{\delta V(4)}d(3) = 0,$$

which implies that

$$\frac{\delta G(12)}{\delta V(3)} = -\int G(14)\frac{\delta G^{-1}(45)}{\delta V(3)}G(52)\,d(45).$$

In order to calculate the polarization function we shall need the *vertex function*,

$$\Gamma(123) = -\frac{\delta G^{-1}(12)}{\delta V(3)}, \qquad (2.219)$$

which derives its name from the graphical representation of Green's functions by Feynman diagrams. We find

$$\Pi\left(12\right) = \pm i \int G\left(13\right) G\left(41\right) \Gamma\left(342\right) d\left(34\right). \tag{2.220}$$

In order to make use of this formula we need an equation for the vertex function. A variation of (2.213) leads to

$$\frac{\delta G^{-1}\left(12\right)}{\delta V\left(3\right)} = -\delta\left(12\right)\delta\left(13\right) - \frac{\delta\Sigma'\left(12\right)}{\delta V\left(3\right)},$$

and

$$\frac{\delta\Sigma'\left(12\right)}{\delta V\left(3\right)} = \int \frac{\delta\Sigma'\left(12\right)}{\delta G\left(45\right)}\frac{\delta G\left(45\right)}{\delta V\left(3\right)} d\left(45\right).$$

Using these in (2.219) gives

$$\Gamma\left(123\right) = \delta\left(12\right)\delta\left(13\right) + \int \frac{\delta\Sigma'\left(12\right)}{\delta G\left(45\right)}\frac{\delta G\left(45\right)}{\delta V\left(3\right)} d\left(45\right).$$

Since, as was shown above,

$$\frac{\delta G\left(12\right)}{\delta V\left(3\right)} = \int G\left(14\right)\Gamma\left(453\right)G\left(52\right) d\left(45\right),$$

we finally obtain the integral equation

$$\Gamma\left(123\right) = \delta\left(12\right)\delta\left(13\right) + \int \frac{\delta\Sigma'\left(12\right)}{\delta G\left(45\right)} G\left(46\right) G\left(75\right)\Gamma\left(673\right) d\left(4567\right) \tag{2.221}$$

for the vertex function.

We still need an equation for the exchange–correlation self–energy (2.214). Using the expression (2.70) for the self–energy and the variational representation (2.150) for the two–particle propagator, we have

$$\Sigma'\left(12\right) = -i \int \Phi\left(13\right)\frac{\delta G\left(14\right)}{\delta\mu\left(3\right)} G^{-1}\left(42\right) d\left(34\right). \tag{2.222}$$

By varying (2.69) with respect to $\mu(1)$, we find

$$\int \frac{\delta G\left(13\right)}{\delta\mu\left(4\right)} G^{-1}\left(32\right) d\left(3\right) + \int G\left(13\right)\frac{\delta G^{-1}\left(32\right)}{\delta\mu\left(4\right)} d\left(3\right) = 0.$$

Therefore (2.222) can be rewritten as

$$\Sigma'\left(12\right) = i \int \Phi\left(13\right) G\left(14\right)\frac{\delta G^{-1}\left(42\right)}{\delta\mu\left(3\right)} d\left(34\right). \tag{2.223}$$

Using the relation

$$\frac{\delta G^{-1}(12)}{\delta \mu(3)} = \int \frac{\delta G^{-1}(12)}{\delta V(4)} \frac{\delta V(4)}{\delta \mu(3)} d(4),$$

and the definitions (2.216) and (2.219), the equation (2.223) takes the form

$$\Sigma'(12) = i \int G(13) \Gamma(324) W(41) d(34). \qquad (2.224)$$

We have thus obtained a closed system of five equations (2.215), (2.218), (2.220), (2.221), and (2.224) for $W, \varepsilon^{-1}, \Pi, \Gamma$, and $\Sigma'$, respectively. The equations (2.215) and (2.218) may be combined to yield the equation

$$W(12) = \Phi(12) + \int \Phi(13) \Pi(34) W(42) d(34) \qquad (2.225)$$

for the screened potential.

To provide a conspectus of this set of equations we write them in symboolic form as follows:

$$W = \varepsilon^{-1}\Phi,$$
$$\varepsilon^{-1} = 1 + \Phi\Pi\varepsilon^{-1},$$
$$\Pi = \pm iGG\Gamma, \qquad (2.226)$$
$$\Gamma = 1 + \frac{\delta\Sigma'}{\delta G}GG\Gamma,$$
$$\Sigma' = iG\Gamma W.$$

This system of equations may be solved by an iteration procedure, starting from the Hartree approximation corresponding to

$$\Sigma'_0(12) = 0. \qquad (2.227)$$

Substituting this into the right–hand side of (2.226) gives

$$\Gamma_0(123) = \delta(12)\,\delta(13) \qquad (2.228)$$

and hence

$$\Pi_0(12) = \pm iG(12)G(21). \qquad (2.229)$$

Let $\varepsilon^{-1}(12) = \delta(12)$, then

$$W_0(12) = \Phi(12). \qquad (2.230)$$

This implies that screening is absent in the zero approximation and the second equation of (2.226) gives the inverse dielectric function

$$\varepsilon_{HF}^{-1}(12) = \delta(12) + \int \Phi(13) \Pi_0(32) d(3) \qquad (2.231)$$

in the Hartree–Fock approximation in which the exchange–correlation self-energy is

$$\Sigma'_{HF}(12) = iG(12)\,\Phi(12). \tag{2.232}$$

Since we began with Hartree approximation, the propagators in (2.231) and (2.232) are also to be taken in this approximation.

Substituting (2.229) into the second equation (2.226) gives the equation

$$\varepsilon_{RP}^{-1}(12) = \delta(12) + \int \Phi(13)\,\Pi_0(34)\,\varepsilon_{RP}^{-1}(42)\,d(34), \tag{2.233}$$

the solution of which is the inverse dielectric function in the *random–phase approximation*. The corresponding screened potential is

$$W_{RP}(12) = \int \varepsilon_{RP}^{-1}(13)\,\Phi(32)\,d(3). \tag{2.234}$$

Inserting this into the fifth equation (2.226) defines the exchange–correlation self-energy

$$\Sigma'_{SP}(12) = iG(12)W_{RP}(12) \tag{2.235}$$

in the *screened–potential approximation.*

The iteration procedure can be continued employing either (2.232) or (2.235). However, successive approximations soon become overly complicated. In order to avoid excessive complications, it is sometimes more convenient to try to include in the zero approxiamtion more information about the properites of the system. For example, instead of starting with the Hartree approximation (2.227), we might begin with the *local–density approximation*

$$\Sigma'_{LD}(12) = V_{LD}(1)\,\delta(12), \tag{2.236}$$

in which the *local–density potential*

$$V_{LD}(1) = V_{LD}\left[\rho(1)\right] \tag{2.237}$$

is a *functional* of the local density $\rho(1)$ such that (2.237) accounts for part of the exchange and correlation effects. Because of this, (2.237) is also called the *local exchange–correlation potential.*

Variation of (2.236) with respect to the propagator $G$ gives

$$\frac{\delta\Sigma'_{LD}(12)}{\delta G(34)} = \pm i\delta(12)\,K_{LD}(23)\,\delta(34),$$

where

$$K_{LD}(12) := \frac{\delta V_{LD}(1)}{\delta\rho(2)}. \tag{2.238}$$

It then follows by means of (2.226) that we can express the vertex function in the form

$$\Gamma(123) = \delta(12)\,\Gamma_{LD}(23), \tag{2.239}$$

where the new function, $\Gamma_{LD}(12)$, called the *local–density vertex,* satisfies the equation

$$\Gamma_{LD}(12) = \delta(12) + \int K_{LD}(13)\,\Pi_0(34)\,\Gamma_{LD}(42)\,d(34). \qquad (2.240)$$

Substituting for $\Gamma(123)$ in the third equation (2.226) leads to the following formula for the polarization function

$$\Pi_{LD}(12) = \int \Pi_0(13)\,\Gamma_{LD}(32)\,d(3) \qquad (2.241)$$

in the local–density approximation. In this approximation, the inverse dielectric function is given by a solution of the integral equation

$$\varepsilon_{LD}^{-1}(12) = \delta(12) + \int \Phi(13)\,\Pi_0(34)\,\Gamma_{LD}(45)\varepsilon_{LD}^{-1}(52)d(345). \qquad (2.242)$$

Consequently, the screened potential in the local–density approximation is

$$W_{LD}(12) = \int \varepsilon_{LD}^{-1}(13)\,\Phi(32)\,d(3). \qquad (2.243)$$

Substituting this in (2.226), we define the screened potential

$$W_{LV}(12) := \int \Gamma_{LD}(13)W_{LD}(32)d(3) \qquad (2.244)$$

in the *local–vertex approximation.* It follows from (2.240) that the potential (2.244) satisfies the equation

$$W_{LV}(12) = \int \Gamma_{LD}(13)\,\Phi(32)\,d(3) +$$

$$+ \int \Gamma_{LD}(13)\,\Phi(34)\,\Pi_0(45)\,W_{LV}(52)\,d(345).$$

With a solution of this equation we may define the local–vertex approximation,

$$\Sigma'_{LV}(12) = iG(12)\,W_{LV}(21), \qquad (2.245)$$

for the exchange–correlation self–energy [65–69].

Since we started from the local–density approximation (2.236), we should take the propagators in (2.244) and (2.245) in the same approximation.

Analysing the structure of perturbation series in quantum electrodynamics, Feynman suggested a graphycal interpretation of perturbative terms in the form of diagrams. Such an interpretation permitted to understand better the physical meaning of perturbative expansions. It helped also to separate an infinite class of diagrams which could be summed, thus, renormalizing and improving perturbation theory. This resummation is usually done by selecting those diagrams which correspond to a set of terms forming a geometrical progression. Together with the penetration of the field–theory methods into the many–body problem, the diagram technique has become popular among the statistical–physics community [70].

We shall not consider here such diagrams because of the following: A diagram is nothing but the illustration of an analytical expression. It is possible to accomplish calculations without diagrams, while, dealing with diagrams, one always has to return finally to analytical formulas. In many cases the diagrammatic illustration helps to clarify the resummation procedure. However, the latter can always be done without invoking the former.

## 2.10   Heterophase States

Till now, speaking about a statistical system we have tacitly assumed that the same type of order occurs in the whole volume of the system. It may happen, however, that different types of order appear in different parts of the system. If these regions of different order are macroscopic then one says that the system is divided into domains. In this case, each domain plays the role of a separate system, so that the problem is actually reduced to that of a system with one fixed order, with a slight complication coming from the boundary conditions describing interdomain layers.

A much more complex situation develops if the regions of different order are of mesoscopic sizes and, moreover, are randomly distributed in space. Since each type of order is related to a thermodynamic phase, we may say then that the system is heterophase. The mesoscopic regions of different phases not only are randomly distributed in space but may also oscillate in time. Because of this, they are called *heterophase fluctuations*. To describe the system with such fluctuations is a problem far from trivial. In this section we present a general approach for treating statistical systems with heterophase fluctuations [19,71–80].

Consider several thermodynamic phases which we enumerate by the index $\nu = 1, 2, \ldots$. Recall that different thermodynamic phases correspond to qualitatively different statistical states. At the same time, to each phase one can put into correspondence a *space of typical states* which is a Hilbert space $\mathcal{H}_\nu$ of microscopic states $h_\nu \in \mathcal{H}_\nu$ such that one can define an *order operator* $\hat{\eta}$ whose mathematical expectation

$$(h_\nu, \hat{\eta} h_\nu) = \eta(h_\nu)$$

gives an *order parameter* $\eta(h_\nu)$ which is qualitatively different from $\eta(h_\mu)$ if $\mu \neq \nu$.

The direct sum

$$\mathcal{X} = \oplus_\nu \mathcal{H}_\nu \tag{2.246}$$

describes a system which can be in either one or another thermodynamic phase. In order to describe the states of coexisting phases we need to deal with the space

$$\mathcal{Y} = \otimes_\nu \mathcal{H}_\nu \tag{2.247}$$

which is the tenzor product of $\mathcal{H}_\nu$. The spaces $\mathcal{X}$ and $\mathcal{Y}$ are uniquely isomorphic but correspond to different physical situations.

Let a statistical system in the real–space region $\mathbf{V} \subset \mathbf{R}^3$ be divided into subregions $\mathbf{V}_\nu \subset \mathbf{V}$ filled by different thermodynamic phases. A family $\{\mathbf{V}_\nu\}$ of subregions $\mathbf{V}_\nu$ forms a covering of the region $\mathbf{V}$ when

$$\mathbf{V} = \cup_\nu \mathbf{V}_\nu, \qquad V = \sum_\nu V_\nu, \tag{2.248}$$

where

$$V = mes\mathbf{V}, \qquad V_\nu = mes\mathbf{V}_\nu.$$

The space separation of subregions $\mathbf{V}_\nu$ can be characterized by the *manifold indicator functions*

$$\xi_\nu(\vec{r}) := \begin{cases} 1, & \vec{r} \in \mathbf{V}_\nu \\ 0, & \vec{r} \notin \mathbf{V}_\nu. \end{cases} \tag{2.249}$$

A set

$$\xi := \{\xi_\nu(\vec{r}) \mid \vec{r} \in \mathbf{V}, \nu = 1, 2, \ldots\} \tag{2.250}$$

of the indicator functions (2.249) uniquely defines a *phase configuration* in real space. Operators of the density observables acting on $\mathcal{H}_\nu$ have the general form

$$\hat{A}_\nu(\xi, \vec{r}) = \sum_{k=0}^\infty \int \xi_\nu(\vec{r}) \xi_\nu(\vec{r}_1) \xi_\nu(\vec{r}_2) \ldots \xi_\nu(\vec{r}_k) \times$$

$$\times A_{\nu k}(\vec{r}, \vec{r}_1, \vec{r}_2, \ldots, \vec{r}_k) d\vec{r}_1 d\vec{r}_2 \ldots d\vec{r}_k, \tag{2.251}$$

in which $A_{\nu k}(\ldots)$ is an operator distribution. Here and in what follows the integrals in real space are assumed to be over the whole region $\mathbf{V}$, if not specified. Operators of local observables, for a subregion $\Lambda$, are given by

$$A_\nu(\xi, \Lambda) = \int_\Lambda \hat{A}_\nu(\xi, \vec{r}) d\vec{r}. \tag{2.252}$$

The operator density (2.251) can be considered as a limiting case of (2.252) in the sense of the limit

$$\lim_{mes\Lambda \to 0} \frac{A_\nu(\xi, \Lambda)}{mes\Lambda} = \hat{A}_\nu(\xi, \vec{r}).$$

The local operators acting on the space (2.247) are defined as

$$\hat{A}(\xi, \vec{r}) = \oplus_\nu \hat{A}_\nu(\xi, \vec{r}), \qquad A(\xi, \Lambda) = \oplus_\nu A_\nu(\xi, \Lambda). \tag{2.253}$$

The family

$$\mathcal{A}(\xi) := \{\hat{A}(\xi, \vec{r}), A(\xi, \Lambda) \mid \vec{r} \in \mathbf{V}, \Lambda \subset \mathbf{V}\} \tag{2.254}$$

of all operators of local observables forms the algebra of local observables.

When a phase configuration characterized by the set (2.250) is fixed, then we have the case of frozen phase separation. However, when the phase configuration is randomly distributed in space and, in addition, may fluctuate in time, we have to consider the set

(2.250) as a stochastic variable. Then observable quantities are given by the averages of operators from the algebra of local observables (2.254), with the averaging procedure containing the trace over microscopic degrees of freedom and, also, an averaging over phase configurations. The latter is a functional integral over the manifold indicator functions (2.249), which we shall denote by $\int \mathcal{D}\xi$. The statistical averaging, as usual, must include a statistical operator $\rho(\xi)$. Thus, the observable quantities are given by the average

$$\langle \mathcal{A} \rangle = Tr \int \rho(\xi)\mathcal{A}(\xi)\mathcal{D}\xi. \tag{2.255}$$

Here and everywhere in what follows it is assumed that the operators of taking trace and of integrating over $\xi$ always commute.

As examples of the operators (2.253), we may write the operator of the energy density

$$\hat{E}(\xi, \vec{r}) = \oplus_\nu \hat{E}_\nu(\xi, \vec{r}), \tag{2.256}$$

where

$$\hat{E}_\nu(\xi, \vec{r}) = \hat{K}_\nu(\xi, \vec{r}) + \hat{V}_\nu(\xi, \vec{r}) \tag{2.257}$$

is the operator of the energy density for a $\nu$–phase, consisting of the kinetic–energy operator density.

$$\hat{K}_\nu(\xi, \vec{r}) = \xi_\nu(\vec{r})\psi_\nu^\dagger(\vec{r}) \left[ -\frac{\nabla^2}{2m} + U(\vec{r}) \right] \psi_\nu(\vec{r}) \tag{2.258}$$

and of the potential–energy operator density

$$\hat{V}_\nu(\xi, \vec{r}) = \frac{1}{2} \int \xi_\nu(\vec{r})\xi_\nu(\vec{r}\,')\psi_\nu^\dagger(\vec{r})\psi_\nu^\dagger(\vec{r}\,') \times$$

$$\times \Phi(\vec{r} - \vec{r}\,')\psi_\nu(\vec{r}\,')\psi_\nu(\vec{r})d\,\vec{r}\,'. \tag{2.259}$$

Another example is the operator of particle density

$$\hat{N}(\xi, \vec{r}) = \oplus_\nu \hat{N}_\nu(\xi, \vec{r}) \tag{2.260}$$

with

$$\hat{N}_\nu(\xi, \vec{r}) = \xi_\nu(\vec{r})\psi_\nu^\dagger(\vec{r})\psi_\nu(\vec{r}). \tag{2.261}$$

To calculate the average (2.255), we need to know the statistical operator $\rho(\xi)$. This can be defined by invoking the *principle of unbiased guess*, that is, by maximizing the information entropy. The entropy itself is defined by the expression

$$S(\rho) = -Tr \int \rho(\xi) \ln \rho(\xi)\mathcal{D}\xi. \tag{2.262}$$

As additional conditions for $\rho(\xi)$, we require that this statistical operator is normalized so that

$$Tr \int \rho(\xi)\mathcal{D}\xi = 1, \tag{2.263}$$

that the local energy density is given by

$$E(\vec{r}) = Tr \int \rho(\xi)\hat{E}(\xi, \vec{r})\mathcal{D}\xi, \tag{2.264}$$

and that the local particle density is

$$N(\vec{r}) = Tr \int \rho(\xi)\hat{N}(\xi, \vec{r})\mathcal{D}\xi. \tag{2.265}$$

For the informational entropy we have

$$S_{inf} = S(\rho) + \zeta \left[ Tr \int \rho(\xi)\mathcal{D}\xi - 1 \right] +$$

$$+ \int \beta(\vec{r}) \left[ E(\vec{r}) - Tr \int \rho(\xi)\hat{E}(\xi, \vec{r})\mathcal{D}\xi \right] d\vec{r} +$$

$$+ \int \gamma(\vec{r}) \left[ N(\vec{r}) - Tr \int \rho(\xi)\hat{N}(\xi, \vec{r})\mathcal{D}\xi \right] d\vec{r}, \tag{2.266}$$

where $\zeta$, $\beta(\vec{r})$, and $\gamma(\vec{r})$ are the Lagrange multipliers. Conditions (2.263), (2.264), and (2.265) are recovered from the variational equations

$$\frac{\partial S_{inf}}{\partial \zeta} = 0, \qquad \frac{\delta S_{inf}}{\delta \beta(\vec{r})} = 0, \qquad \frac{\delta S_{inf}}{\delta \gamma(\vec{r})} = 0.$$

The statistical operator $\rho(\xi)$ is defined as a maximizer of (2.266), that is, from the equations

$$\frac{\delta S_{inf}}{\delta \rho(\xi)} = 0, \qquad \frac{\delta^2 S_{inf}}{\delta \rho^2(\xi)} < 0. \tag{2.267}$$

Introducing the notation

$$\zeta = 1 - \ln Z, \qquad \gamma(\vec{r}) = -\beta(\vec{r})\mu(\vec{r}),$$

from the extremum condition of (2.267) we obtain

$$\rho(\xi) = \frac{1}{Z} \exp\left\{-Q(\xi)\right\}, \tag{2.268}$$

where

$$Z = Tr \int \exp\{-Q(\xi)\}\mathcal{D}\xi \tag{2.269}$$

is the partition function and

$$Q(\xi) = \int \beta(\vec{r}) \left[ \hat{E}(\xi, \vec{r}) - \mu(\vec{r})\hat{N}(\xi, \vec{r}) \right] d\vec{r} \tag{2.270}$$

is the *quasi–Hamiltonian*.

Now we need to concretize the procedure of averaging over phase configurations, that is, we have to define explicitly the functional integration over the stochastic variable (2.250). All conceivable variants of the phase configurations form a topological space $\mathcal{K} = \{\xi\}$. To integrate over the stochastic variable $\xi$, we have to define a functional measure $\int \mathcal{D}\xi$ on the space $\mathcal{K}$.

Note that averaging over space configurations contains two types of actions. The first one averages over all possible phase configurations under a fixed set

$$x := \{x_\nu | \; \nu = 1, 2, \ldots\} \tag{2.271}$$

of geometric probabilities

$$x_\nu := \frac{1}{V} \int \xi_\nu(\vec{r}) d\vec{r} = \frac{V_\nu}{V} \tag{2.272}$$

with the evident properties

$$\sum_\nu x_\nu = 1, \qquad 0 \le x_\nu \le 1. \tag{2.273}$$

The second action is the variation of each probability (2.272) from zero to unity taking account of their normalization (2.273).

The resulting differential functional measure $\mathcal{D}\xi$ can be written as

$$\mathcal{D}\xi = dx \mathcal{D}_x \xi \tag{2.274}$$

with

$$dx = \delta \left( \sum_\nu x_\nu - 1 \right) \prod_\nu dx_\nu, \tag{2.275}$$

where $x_\nu \in [0, 1]$.

To define $\mathcal{D}_x \xi$, let us divide the whole system into $n$ parts, so that each subregion $\mathbf{V}_\nu$ is covered by a family of $n_\nu$–subregions $\mathbf{V}_{\nu i}$ such that

$$\mathbf{V}_\nu = \bigcup_{i=1}^{n_\nu} \mathbf{V}_{\nu i}, \qquad V_\nu = \sum_{i=1}^{n_\nu} v_{\nu i}$$

and with

$$n = \sum_\nu n_\nu, \qquad v_{\nu i} = mes\mathbf{V}_{\nu i}.$$

The shape of each sub–region $\mathbf{V}_{\nu i}$ can be arbitrary. For each $\mathbf{V}_{\nu i}$, let us fix a vector $\vec{a}_{\nu i} \in \mathbf{V}_{\nu i}$ defined as the center of the local coordinate system in $\mathbf{V}_{\nu i}$. Then the manifold indicator function (2.249) can be presented as

$$\xi_\nu(\vec{r}) = \sum_{i=1}^{n_\nu} \xi_{\nu i}(\vec{r} - \vec{a}_{\nu i}), \tag{2.276}$$

where

$$\xi_{\nu i}(\vec{r} - \vec{a}_{\nu i}) = \begin{cases} 1, & \vec{r} \in \mathbf{V}_{\nu i} \\ \\ 0, & \vec{r} \notin \mathbf{V}_{\nu i}. \end{cases}$$

It is clear that if the sub–subregions $\mathbf{V}_{\nu i}$ are small enough, then they can be used as blocks for producing any phase configuration. A displacement of the blocks can be accomplished by moving the block centers $\vec{a}_{\nu i}$.

In this way, we come to the definition of the functional differential

$$\mathcal{D}_x \xi := \prod_\nu \prod_{i=1}^{n_\nu} \frac{a_{\nu i}}{V} \qquad (n \to \infty), \tag{2.277}$$

in which $\vec{a}_{\nu i} \in \mathbf{V}$ and the limit $n \to \infty$ implies that

$$n \to \infty : \ n_\nu \to \infty, \quad v_{\nu i} \to 0, \quad x_\nu \to const. \tag{2.278}$$

The limit (2.278) is to be taken after the corresponding functional integration. Thus, for a function of $\xi$ denoted by $F(\xi)$ we have

$$\int F(\xi)\mathcal{D}_x \xi = \lim_{n \to \infty} \int F(\xi) \prod_\nu \prod_{i=1}^{n_\nu} \frac{d \vec{a}_{\nu i}}{V}. \tag{2.279}$$

For example, it is straightforward to check that

$$\int \mathcal{D}_x \xi = 1, \qquad \int \xi_\nu(\vec{r})\mathcal{D}_x \xi = x_\nu.$$

More generally, the following statement holds.

**Proposition 2.1.**

Consider a class of functionals of the form

$$F(\xi) = \sum_{k=0}^{\infty} \sum_{\nu_1} \sum_{\nu_2} \cdots \sum_{\nu_k} \int \xi_{\nu_1}(\vec{r}_1)\xi_{\nu_2}(\vec{r}_2) \dots \xi_{\nu_k}(\vec{r}_k) \times$$

$$\times F_{\nu_1 \nu_2 \dots \nu_k}(\vec{r}_1, \vec{r}_2, \dots, \vec{r}_k) d\vec{r}_1 d\vec{r}_2 \dots d\vec{r}_k. \tag{2.280}$$

Then the functional integration (2.279) gives

$$\int F(\xi)\mathcal{D}_x \xi = F(x), \tag{2.281}$$

where $F(x)$ has the form of (2.280) but with all $\xi_\nu(\vec{r})$ changed by $x_\nu$.

**Proof** is based on a direct integration of products of the manifold indicator functions (2.276) over the block vectors $\vec{a}_{\nu i}$ with the differential measure (2.277). In this integration, the products of $\xi_{\nu i}$ with coinciding vectors $\vec{a}_{\nu i}$ yield Lesbegue zeros. For example,

$$\sum_{i=1}^{n_\nu} \int \xi_{\nu i}(\vec{r}_1 - \vec{a}_{\nu i})\xi_{\nu i}(\vec{r}_2 - \vec{a}_{\nu i})\mathcal{D}_x \xi =$$

$$= \lim_{n \to \infty} \sum_{i=1}^{n_\nu} \int \xi_{\nu i}(\vec{r}_1 - \vec{a}_{\nu i}) \xi_{\nu i}(\vec{r}_2 - \vec{a}_{\nu i}) \frac{d\,\vec{a}_{\nu i}}{V} = x_\nu \delta_{r_1 r_2},$$

where $\delta_{r_1 r_2}$ is the Kronecker delta. Therefore, only the manifold indicator functions with different block vectors $\vec{a}_{\nu i}$ contribute to (2.279) which leads to (2.281).

**Proposition 2.2.**

The partition function (2.269) with the quasi–Hamiltonian (2.270) and the differential functional measure (2.274) can be written as

$$Z = Tr \int \exp\{-Q(x)\} dx, \tag{2.282}$$

where $Q(x)$ has the form of (2.270) but with all $\xi_\nu(\vec{r})$ changed by $x_\nu$, and $dx$ is defined in (2.275).

**Proof** involves the previous proposition after expanding the exponential in (2.269) in powers of $Q(\xi)$ and obtaining

$$\int \exp\{-Q(\xi)\} \mathcal{D}_x \xi = \exp\{-Q(x)\}.$$

The integration in (2.282) with the differential measure (2.275), if accomplished in the thermodynamic limit $N \to \infty$, defines a natural thermodynamic potential for a heterophase system as follows.

**Proposition 2.3.**

Let a function

$$y(x) := - \lim_{N \to \infty} \frac{1}{N} \ln Tr \exp\{-Q(x)\} \tag{2.283}$$

of $x$ in the hypercube

$$\mathbf{H} := \{x|\ \sum_\nu x_\nu = 1,\ 0 \le x_\nu \le 1\}$$

have an absolute minimum

$$y(w) = abs \min_{x \in \mathbf{H}} y(x), \tag{2.284}$$

in which the set

$$w := \{w_\nu|\ \nu = 1, 2 \ldots\} \tag{2.285}$$

consists of elements with the property

$$\sum_\nu w_\nu = 1, \qquad 0 \le w_\nu \le 1. \tag{2.286}$$

Then, in the thermodynamic limit,

$$y(w) = - \lim_{N \to \infty} \frac{1}{N} \ln Z. \tag{2.287}$$

**Proof** uses the asymptotic, as $N \to \infty$, presentation

$$Z \simeq \int \exp\{-N y(x)\} dx$$

for the partition function (2.282) with definition (2.283). Then, the Laplace method of integration immediately yields (2.287).

Thus, we come to the conclusion that the role of a natural thermodynamic potential for a heterophase system is played by the function

$$y(w) = -\lim_{N\to\infty} \frac{1}{N} \ln Tr \exp\{-Q(w)\}. \tag{2.288}$$

The set (2.285) consists of the geometric probabilities $w_\nu$ defining relative weights of each thermodynamic phase. Consequently, $w_\nu$ may be called the *phase probability*. To find these probabilities, we have to minimize the *heterophase thermodynamic potential* (2.288). The minimization means that the probabilities $w_\nu$ are to be defined by the equations

$$\frac{\partial}{\partial w_\nu} y(w) = 0, \qquad \frac{\partial^2}{\partial w_\nu^2} y(w) > 0 \tag{2.289}$$

under condition (2.286). In addition, one has to analyse the thermodynamic potential $y(w)$ for $w$ on the boundary of the hupercube $\mathbf{H}$, that is, when one or several $w_\nu$ are equal to one or zero. The first equation in (2.289) can be called the *equation for phase probabilities*, and the inequality in (2.289) is the *condition of heterophase stability*.

Taking account of the form of the heterophase thermodynamic potential (2.288), as the equation for phase probabilities we get

$$\lim_{N\to\infty} \frac{1}{N} \left\langle \frac{\partial}{\partial w_\nu} Q(w) \right\rangle = 0. \tag{2.290}$$

And the condition of heterophase stability becomes

$$\lim_{N\to\infty} \frac{1}{N} \left[ \left\langle \frac{\partial^2}{\partial w_\nu^2} Q(w) \right\rangle - \left\langle \frac{\partial}{\partial w_\nu} Q(w) \right\rangle^2 \right] > 0. \tag{2.291}$$

The quasi–Hamiltonian $Q(w)$ is defined in (2.270) which together with (2.256)–(2.261) gives

$$Q(w) = \oplus_\nu Q_\nu(w), \tag{2.292}$$

where

$$Q_\nu(w) = w_\nu \int \beta(\vec{r}) \psi_\nu^\dagger(\vec{r}) \left[ -\frac{\nabla^2}{2m} + U(\vec{r}) - \mu(\vec{r}) \right] \psi_\nu(\vec{r}) d\vec{r} +$$

$$+ \frac{1}{2} w_\nu^2 \int \beta(\vec{r}) \psi_\nu^\dagger(\vec{r}) \psi_\nu^\dagger(\vec{r}') \Phi(\vec{r} - \vec{r}') \psi_\nu(\vec{r}') \psi_\nu(\vec{r}) d\vec{r} \, d\vec{r}' . \tag{2.293}$$

Finally, we need to find an effective presentation for the expectation values (2.255) of the algebra of local observables (2.254).

**Proposition 2.4.**

Let the conditions of the Proposition 2.3 be valid, then the expectation values (2.255) asymptotically, as $N \to \infty$, take the form

$$\langle \mathcal{A} \rangle \simeq Tr \rho_{eff}(w) \mathcal{A}(w) \qquad (N \to \infty), \qquad (2.294)$$

where $\mathcal{A}(w)$ is the algebra of local observables (2.254) with all $\xi_\nu(\vec{r})$ changed by $w_\nu$, and the *effective statistical operator* is

$$\rho_{eff}(w) := \frac{1}{Z_{eff}} \exp\{-Q(w)\} \qquad (2.295)$$

with the *effective partition function*

$$Z_{eff} := Tr \exp\{-Q(w)\}. \qquad (2.296)$$

**Proof** starts from Proposition 2.1, according to which

$$\int \rho(\xi) \mathcal{A}(\xi) \mathcal{D}_x \xi = \rho(x) \mathcal{A}(x),$$

where

$$\rho(x) = \frac{1}{Z} \exp\{-Q(x)\}$$

with $Z$ from (2.282). Then, introducing

$$\bar{A}(x) := \frac{Tr \exp\{-Q(x)\} \mathcal{A}(x)}{\exp\{-Ny(x)\}}, \qquad (2.297)$$

we have

$$Tr \rho(x) \mathcal{A}(x) = \frac{1}{Z} \exp\{-Ny(x)\} \bar{A}(x).$$

Using the method of steepest descent, for $N \to \infty$, we obtain

$$\int \exp\{-Ny(x)\} \bar{A}(x) dx \simeq \exp\{-Ny(w)\} \bar{A}(w) \prod_\nu \left(\frac{2\pi}{Ny_\nu''}\right)^{1/2},$$

where the product over $\nu$ contains the number of factors by one less than the number of phases, since one of $w_\nu$ must be expressed through others because of the normalization (2.286), and

$$y_\nu'' \equiv \frac{\partial^2 y(w)}{\partial w_\nu^2} > 0.$$

Similarly, we find

$$Z \simeq \exp\{-Ny(w)\} \prod_\nu \left(\frac{2\pi}{Ny_\nu''}\right)^{1/2}.$$

It follows that

$$\frac{1}{Z} \int \exp\{-N y(x)\} \bar{A}(x) dx \simeq \bar{A}(w),$$

as $N \to \infty$, that is

$$\langle A \rangle \simeq \bar{A}(w).$$

Remembering the definition (2.297), we come to (2.294) with (2.295) and (2.296).

When there are no external fields specially inducing nonuniformity, the local inverse temperature and chemical potential are to be uniform through the system:

$$\beta(\vec{r}) = \beta, \qquad \mu(\vec{r}) = \mu. \tag{2.298}$$

These equations have the standard form of equilibrium conditions. However, we should not forget that a heterophase system, as is explained above, is quasiequilibrium. Condition (2.298) appears after averaging over phase configurations. Therefore this condition has the meaning of the *condition of equilibrium on average* or it may be called the *condition of heterophase equilibrium.*

Under condition (2.298), the quasi–Hamiltonian (2.292) becomes

$$Q(w) = \beta H(w) \tag{2.299}$$

with the *renormalized* or *effective heterophase Hamiltonian*

$$H(w) := \oplus_\nu H_\nu(w) \tag{2.300}$$

consisting of the terms

$$H_\nu(w) = w_\nu \int \psi_\nu^\dagger(\vec{r}) \left[ -\frac{\nabla^2}{2m} + U(\vec{r}) - \mu \right] \psi_\nu(\vec{r}) d\vec{r} +$$

$$+ \frac{1}{2} w_\nu^2 \int \psi_\nu^\dagger(\vec{r}) \psi_\nu^\dagger(\vec{r}\,') \Phi(\vec{r} - \vec{r}\,') \psi_\nu(\vec{r}\,') \psi_\nu(\vec{r}) d\vec{r} \, d\vec{r}\,'. \tag{2.301}$$

Emphasize that (2.300) represents not just one system with a given phase separation but an infinite number of systems with all possible phase configurations. Because of this, each term (2.301) may be called the *phase–replica Hamiltonian.* Owing to the nonlinear renormalization in (2.301), interphase effects are included into the description. The *interphase state* is defined as

$$\langle A \rangle_{int} := \langle A \rangle - \sum_\nu w_\nu \langle A \rangle_\nu, \tag{2.302}$$

that is, as the difference between the heterophase state $\langle A \rangle$, and the respectively weighted sum of pure states

$$\langle A \rangle_\nu := \lim_{w_\nu \to 1} \langle A \rangle.$$

As an example, consider two coexisting phases. Then the phase–probability equation (2.290), with the quasi–hamiltonian (2.299) gives

$$w_1 = \frac{2\Phi_2 + K_2 - K_1 + \mu(R_1 - R_2)}{2(\Phi_1 + \Phi_2)}, \qquad w_2 = 1 - w_1, \tag{2.303}$$

where

$$K_\nu = \frac{1}{N} \int \left\langle \psi_\nu^\dagger(\vec{r}) \left[ -\frac{\nabla^2}{2m} + U(\vec{r}) \right] \psi_\nu(\vec{r}) \right\rangle d\vec{r},$$

$$\Phi_\nu = \frac{1}{N} \int \left\langle \psi_\nu^\dagger(\vec{r}) \psi_\nu^\dagger(\vec{r}^{\,\prime}) \Phi(\vec{r} - \vec{r}^{\,\prime}) \psi_\nu(\vec{r}^{\,\prime}) \psi_\nu(\vec{r}) \right\rangle d\vec{r}\, d\vec{r}^{\,\prime},$$

$$R_\nu = \frac{1}{N} \int \left\langle \psi_\nu^\dagger(\vec{r}) \psi_\nu(\vec{r}) \right\rangle d\vec{r}.$$

Thus, all phase probabilities are defined in a self–consistent way.

Each field operator $\psi_\nu(\vec{r})$ is defined on a weighted Hilbert space $\mathcal{H}_\nu$ and has the standard commutation or anticommutation relations,

$$\left[ \psi_\nu(\vec{r}), \psi_\nu(\vec{r}^{\,\prime}) \right]_\mp = 0, \qquad \left[ \psi_\nu(\vec{r}), \psi_\nu^\dagger(\vec{r}^{\,\prime}) \right] = \delta(\vec{r} - \vec{r}^{\,\prime}),$$

depending on the kind of statistics. Field operators defined on different spaces, by definition, commute with each other:

$$\left[ \psi_\nu(\vec{r}), \psi_{\nu'}(\vec{r}^{\,\prime}) \right] = 0 = \left[ \psi_\nu(\vec{r}), \psi_{\nu'}^\dagger(\vec{r}^{\,\prime}) \right] \qquad (\nu \neq \nu') \tag{2.304}$$

Therefore, the evolution equation for

$$\psi_\nu(\vec{r}, t) = U_\nu^+(t) \psi_\nu(\vec{r}) U_\nu(t); \qquad U_\nu)t) = e^{-iH_\nu t},$$

has the usual Heisenberg form

$$i\frac{\partial}{\partial t} \psi_\nu(\vec{r}, t) = \left[ \psi_\nu(\vec{r}, t), H_\nu \right].$$

For each phase replica, indexed by $\nu = 1, 2, \ldots$, we may introduce a propagator

$$G_\nu(\vec{r}, t; \vec{r}^{\,\prime}, t') := -i\langle \hat{T} \psi_\nu(\vec{r}, t) \psi_\nu^\dagger(\vec{r}^{\,\prime}, t') \rangle. \tag{2.305}$$

Similarly, we may introduce many–particle propagators. Then we can write the evolution equation for (2.305) and proceed in analogy with one of the techniques described in the previous sections. Dealing with the effective Hamiltonian (2.300), we have the dynamical equations for different phase replicas formally decoupled, although thermodynamically all equations are related with each other through the geometric probabilities (2.285). The possibility of realizing this dynamical decoupling is one of the main advantages of the approach formulated in the theory of heterophase fluctuations [19,71–80].

# 3  Correlated Iteration

The Green function technique expounded in the previous Chapter can be applied to many statistical systems. However, there exists an important class of systems which exhibits such a strong correlation between particles that a direct application of the methods discussed in the previous Chapter becomes impossible. For such systems there should be developed a special approach taking into account strong interparticle correlations from the very beginning.

In this Chapter, we shall present such an approach to treating strongly correlated systems, based on Refs. [81-84].

## 3.1  Singular Potentials

The iteration procedures for developing a perturbation series which we have discussed hitherto are convenient for systems of particles interacting through non–singular or weakly singular potentials. We make these terms more precise by classifying potentials according to their behaviour as $r_{12} \to 0$, where $r_{12} \equiv |\mathbf{r}_{12}|$, and $\mathbf{r}_{12} = \mathbf{r}_2 - \mathbf{r}_1$. We shall say that $\Phi(x_1, x_2)$ is a *non–singular potential* if

$$\lim_{r_{12} \to 0} |\Phi(x_1, x_2)| < \infty. \tag{3.1}$$

A *singular potential* is one for which

$$\lim_{r_{12} \to 0} |\Phi(x_1, x_2)| = \infty. \tag{3.2}$$

A *weakly singular potential* is singular but is integrable in the sense

$$\left| \int_{\mathcal{R}} \Phi(x_1, x_2) d\mathbf{r}_{12} \right| < \infty , \tag{3.3}$$

where $\mathcal{R}$ is a finite region including $r_{12} = 0$. A *strongly singular potential* is singular and is not integrable so that

$$\left| \int_{\mathcal{R}} \Phi(x_1, x_2) d\mathbf{r}_{12} \right| = \infty. \tag{3.4}$$

For example, if the behaviour of a potential at short distances is governed by the power law

$$\Phi(x_1, x_2) \sim \frac{1}{r_{12}^n} \qquad (r_{12} \to 0) ,$$

and $\mathcal{R}$ is a small finite ball centered at the origin, then the potential, $\Phi$ is nonsingular if $n = 0$, weakly singular if $0 < n \leq 2$, and strongly singular if $n > 2$. Thus the Coulomb–like potentials with $n = 1$, are weakly singular. Whereas the Lennard–Jones potential with $n = 12$, which is sometimes used to model interaction of molecules, is strongly singular.

For systems interacting via a strongly singular potential, any iteration procedure starting from the Hartree–Fock approximation breaks down since in the very first step the Hartree potential is divergent. This is because the input to the Hartree approximation assumes independent particles and does not take into account short–range correlations which could effectively shield the singularity. These short–range correlations might be retained by a decoupling procedure in which the three–particle propagator is approximated by a linear combination of two–particle propagators. However, decoupling at any stage effectively precludes subsequent approximations.

What we really need is a perturbation procedure which satisfies the following conditions: 1) it should start from an approximation containing no divergencies and must therefore take into account strong short–range correlations, 2) it should provide an algorithm which permits recursive calculation of higher–order corrections. We now propose a procedure which we believe meets these desiderata and which we have named the *correlated iteration theory*.

## 3.2   Iterative Algorithm

Let us consider a situation with a strongly singular interaction potential $\Phi(12)$ and let $G_2^0$ be given by (2.75). Suppose we can find a function, $s(12)$, such that

$$G_2^{app}(1221) := -s(12)\rho(1)\rho(2), \qquad t_1 > t_2, \tag{3.5}$$

is an approximation for the 2–particle propagator. We shall call $s(12)$ a *smoothing function*. It follows from this equation that

$$s(12) = s(21). \tag{3.6}$$

We shall require that the smoothed potential

$$\bar{\Phi}(12) := s(12)\Phi(12) \tag{3.7}$$

be nonsingular or weakly singular and therefore integrable in the sense of (3.3).

We now introduce a function, $D(123)$, satisfying the relation

$$G_2(1223) := s(12) \int D(124)G(43)d(4), \tag{3.8}$$

which we call the *doubling function*. This function will be used to transform a 1–particle into a 2–particle propagator, while the smoothing function takes inter–particle correlations into account. An initial approximation, $D_0$, can be obtained by finding a solution of the integral equation obtained by combining (3.5) with (3.8), giving

$$G_2^0(1223) = \int D_0(124)G(43)d(4). \tag{3.9}$$

It is not difficult to verify that

$$D_0(123) := \delta(13)G(22) \pm G(12)\delta(23) \tag{3.10}$$

is a solution of (3.9). Using (3.8), the exact self–energy becomes

$$\Sigma(12) = \pm i \int \bar{\Phi}(13) D(132) d(3) \qquad (3.11)$$

which, in view of (3.10), has as first approximation

$$\Sigma_1(12) = i\bar{\Phi}(12) G(12) + \delta(12) \int \bar{\Phi}(13)\rho(3)d(3). \qquad (3.12)$$

This differs from the Hartree–Fock form in that the bare interaction potential $\Phi$ has been replaced by the smoothed potential $\bar{\Phi}$. Therefore, no divergencies appear in (3.12) which may be called the *first correlated approximation* or the *correlated Hartree–Fock approximation*.

In order to define a recursive procedure for obtaining higher–order approximations we shall need a set of equations which determine the doubling function and the self–energy. Establishing these equations is the next, and most complicated step in our theory. By a variation of the equation of motion (2.69) with respect to $\mu$, we have

$$\frac{\delta G(12)}{\delta\mu(3)} = -\int G(14)\frac{\delta G^{-1}(45)}{\delta\mu(3)} G(52)d(45).$$

From the definition of the inverse propagator it follows that

$$\frac{\delta G^{-1}(12)}{\delta\mu(3)} = \delta(12)\delta(13) - \frac{\delta\Sigma(12)}{\delta\mu(3)},$$

where

$$\frac{\delta\Sigma(12)}{\delta\mu(3)} = \int \frac{\delta\Sigma(12)}{\delta G(45)}\frac{\delta G(45)}{\delta\mu(3)}d(45).$$

Therefore,

$$\frac{\delta G(12)}{\delta\mu(3)} = -G(13)G(32)+$$

$$+ \int G(14)G(52)\frac{\delta\Sigma(45)}{\delta G(67)}\frac{\delta G(67)}{\delta\mu(3)}d(4567).$$

Substituting in this the expression

$$\frac{\delta G(12)}{\delta\mu(3)} = \pm\left[G(12)G(33) - G_2(1332)\right],$$

which results from the variational representation (2.151), we obtain the equation

$$G_2(1223) = G_2^0(1223)+ \qquad (3.13)$$

$$+ \int G(14)\frac{\delta\Sigma(45)}{\delta G(67)}\left[G_2(6227) - G(67)G(22)\right] G(53)d(4567)$$

for the 2–propagator. By using the relation (3.8) which defines the doubling function, we get

$$s(12)D(123) = D_0(123)+ \tag{3.14}$$

$$+ \int G(14)\frac{\delta\Sigma(43)}{\delta G(56)} \left[ s(52)D(527)G(76) - G(22)G(56)\delta(67) \right] d(4567).$$

This suggests that we should define an integral operator $\mathbf{X}$ on $H^3$ as follows

$$\mathbf{X}f(123) := s(12)f(123)+ \tag{3.15}$$

$$+ \int G(14)\frac{\delta\Sigma(43)}{\delta G(56)} \left[ G(22)G(56)\delta(67) - s(52)f(527)G(76) \right] d(4567).$$

Then (3.14) may be rewritten as

$$\mathbf{X}D(123) = D_0(123). \tag{3.16}$$

Now define, as left inverse of $\mathbf{X}$, the operator $\mathbf{Y}$ such that

$$\mathbf{Y} := \mathbf{X}^{-1}, \qquad \mathbf{YX} = 1, \tag{3.17}$$

where $1$ is the identity operator. Acting with $\mathbf{Y}$ on (3.16), we obtain

$$D(123) = \mathbf{Y}D_0(123), \tag{3.18}$$

which is the main equation for the doubling function.

To obtain an iteration procedure for solving (3.18), we proceed as follows. Assume that our initial approximation, (3.5), models the major properties of the system under consideration so that $D$ will be close to $D_0$. In this case, we shall assume that $\mathbf{X}$ is close to the identity. We therefore try to obtain $\mathbf{Y}$ by expressing it as

$$\mathbf{Y} = [1 - (1 - \mathbf{X})]^{-1}, \tag{3.19}$$

which leads to the Neumann series expansion

$$\mathbf{Y} = \sum_{k=0}^{\infty}(1 - \mathbf{X})^k. \tag{3.20}$$

From (3.15) we see that $\mathbf{X}$ is a functional of the self–energy $\Sigma$,

$$\mathbf{X} = \mathbf{X}[\Sigma]. \tag{3.21}$$

Let

$$\mathbf{X}_k = \mathbf{X}[\Sigma_k] \qquad (\mathbf{X}_0 := 1) \tag{3.22}$$

be the same functional acting on the $k$-approximation to the self–energy. Define the $k$-approximation to $\mathbf{Y}$ by

$$\mathbf{Y}_k := \sum_{p=0}^{k}(1 - \mathbf{X}_k)^p \qquad (\mathbf{Y}_0 := 1) . \tag{3.23}$$

If the sequence $\{\mathbf{Y}_k\}$ has a limit, then

$$\mathbf{Y} = \lim_{k\to\infty} \mathbf{Y}_k, \tag{3.24}$$

provided that $\Sigma = \lim_{k\to\infty}\Sigma_k$.

With equations (3.11), (3.18), and (3.20) we have reached our immediate goal of being able to define an iteration process for the calculation of the doubling function and the self–energy. The recursion can be organized according to the scheme:

$$D_k \to \Sigma_{k+1} \to \mathbf{Y}_{k+1} \to D_{k+1}. \tag{3.25}$$

For $k = 0$ we start from (3.10) leading to (3.12). Then because of the variational derivative

$$\frac{\delta\Sigma_1(12)}{\delta G(34)} = i\bar{\Phi}(14)\left[\delta(13)\delta(24) \pm \delta(12)\delta(34)\right],$$

we find

$$\mathbf{Y}_1 = 2 - \mathbf{X}_1. \tag{3.26}$$

It follows that (3.18) gives

$$D_1(123) = [2 - s(12)]D_0(123)\pm$$

$$\pm i \int \left[G(14)G(23) \pm G(13)G(24)\right] G(42)\bar{\Phi}(43)d(4)\mp \tag{3.27}$$

$$\mp i \int \left[G(13)G(44) \pm G(14)G(43)\right] G(22)\left[1 - s(42)\right]\bar{\Phi}(43)d(4).$$

Combining this with (3.11) yields the *second correlated approximation* for the self–energy:

$$\Sigma_2(12) = i\bar{\Phi}(12)G(12) + \delta(12)\int \Phi_2(12)\rho(3)d(3)-$$

$$- \int \bar{\Phi}(13)G_2^0(1324)G(43)\bar{\Phi}(42)d(34)+ \tag{3.28}$$

$$+ \int \Phi(134)G_2^0(1332)G(44)\bar{\Phi}(32)d(34),$$

which involves the following two new functions, the *twice smoothed potential*

$$\Phi_2(12) := \bar{\Phi}(12)[2 - s(12)] = s(12)[2 - s(12)]\Phi(12),$$

and the *effective triple potential*

$$\Phi(123) := \bar{\Phi}(13)[1 - s(23)] = s(13)[1 - s(23)]\Phi(13).$$

Provided that the smoothing function, $s(12)$, is known we may find higher–order approximations by following the scheme (3.25).

113

## 3.3  Smoothing Function

According to (3.5) the smoothing function is

$$s(12) = -\frac{G_2^{app}(1221)}{\rho(1)\rho(2)} \qquad (t_{12} > 0). \tag{3.29}$$

In view of the relations

$$G(11) = \mp i\rho(1), \qquad G(12)G(21) = \mp i\Pi_0(12),$$

the denominator in (3.29) is given by

$$\rho(1)\rho(2) = -G_2^0(1221) - i\Pi_0(12).$$

Since it is clear that the diagonal two–particle propagator, $G_2(1221)$, plays a special role in our correlated approximation procedure, we analyze some of its general properties.

Consider the density–of–particle operator

$$\hat{n}(1) := \psi^\dagger(1)\psi(1), \tag{3.30}$$

which has expected value

$$\rho(1) = \langle \hat{n}(1) \rangle, \tag{3.31}$$

equal to the density–of–particles. It is easy to check that when $t_1 = t_2$,

$$[\hat{n}(1), \hat{n}(2)] = 0.$$

Thus we may define the chronological product

$$\hat{T}\hat{n}(1)\hat{n}(2) := \begin{cases} \hat{n}(1)\hat{n}(2), & t_1 > t_2 \\ \hat{n}(2)\hat{n}(1), & t_1 < t_2 \end{cases}. \tag{3.32}$$

We may then, for the diagonal propagator $G_2$, obtain the expression

$$G_2(1221) = -\langle \hat{T}\hat{n}(1)\hat{n}(2) \rangle. \tag{3.33}$$

Note the symmetry property

$$G_2(1221) = G_2(2112), \tag{3.34}$$

which holds both for $t_1 > t_2$ and $t_1 < t_2$. This is in contrast to the *density–density correlation* function

$$R(12) := \langle \hat{n}(1)\hat{n}(2) \rangle \tag{3.35}$$

for which, in general, $R(12) = R(21)$ only when $t_1 = t_2$.

The correlation function (3.35) is related to the second–order density matrix by the relation

$$\lim_{t_2 \to t_1} R(12) = \delta(x_1 - x_2)\rho(1) + \rho_2(x_1, x_2, x_1, x_2, t_1),$$

which implies that the propagator (3.33) may be expressed in the form

$$G_2(1221) = -\Theta(t_{12})R(12) - \Theta(t_{21})R(21), \tag{3.36}$$

where $t_{12} := t_1 - t_2$.

A quantity which is often considered is the *density–density Green's function* defined by

$$G_\rho(12) := -i\langle \hat{T}\hat{n}(1)\hat{n}(2)\rangle, \tag{3.37}$$

which satisfies spectral representations and dispersion relations similar to those of the Bose propagators studied in Sections 2.2 and 2.3. The propagators (3.36) and (3.37) differ only by a factor $i$,

$$G_\rho(12) = iG_2(1221).$$

Another important quantity is the *pair correlation function*,

$$g(12) := \frac{\langle \hat{n}(1)\hat{n}(2)\rangle}{\langle \hat{n}(1)\rangle\langle \hat{n}(2)\rangle}. \tag{3.38}$$

Since this is related to (3.35) by the equation

$$R(12) = \rho(1)\rho(2)g(12), \tag{3.39}$$

it follows that $g(12)$ and $R(12)$ have the same symmetry properties. In particular,

$$g(12) = g(21), \qquad (t_1 = t_2)\,.$$

By (3.38), the propagator (3.33) can be expressed by the formula

$$G_2(1221) = -\rho(1)\rho(2)\left[\Theta(t_{12})g(12) + \Theta(t_{21})g(21)\right].$$

The equations of motion imply that $G_2$ differs from $G_2^0$ only in that it takes into account interactions between particles. Thus if $\Phi(12)$ approaches zero, then $G_2(1234)$ approaches $G_2^0(1234)$. This implies that the smoothing function (3.29) must satisfy the asymptotic condition

$$s(12) \to 1, \qquad |x_1 - x_2| \to \infty\,. \tag{3.40}$$

The Bethe–Salpeter equation (2.86) will sometimes enable us to find an approximate $G_2^{app}$ needed for the equation (3.29) defining the smoothing function.


## 3.4  Response Functions

The functions discussed in this section enable us to characterize the changes in a system under the action of some external source or when the corresponding macroscopic quantity is varied. For example, the *density response function*

$$\chi(12) := -\frac{\delta\rho(1)}{\delta\mu(2)}, \tag{3.41}$$

describes the change in the average density

$$\delta\rho(1) = -\int \chi(12)\delta\mu(2)$$

when the effective chemical potential is varied.

It is also convenient to introduce the *generalized response function*

$$\chi(123) := \mp i\frac{\delta G(12)}{\delta\mu(3)} \tag{3.42}$$

of which (3.41) is the particular case

$$\chi(12) = \chi(112). \tag{3.43}$$

From the variational representation (2.151) we have

$$\chi(123) = i\left[G_2(1332) - G(12)G(33)\right]. \tag{3.44}$$

By substituting $G_2^0$ we find

$$\chi_0(123) = i\left[G_2^0(1332) - G(12)G(33)\right],$$

which defines in zero–approximation

$$\chi_0(123) = \pm iG(13)G(32) \tag{3.45}$$

for the response function (3.42).

Because of (3.43) the density response function (3.41) can be expressed by

$$\chi(12) = i\left[G_2(1221) + \rho(1)\rho(2)\right]. \tag{3.46}$$

In view of (3.34) this shows that the response function is also symmetric:

$$\chi(12) = \chi(21) = -\frac{\delta\rho(2)}{\delta\mu(1)}. \tag{3.47}$$

An equation for the determination of the generalized response function can be obtained by using (3.44) and (3.14), giving

$$\chi(123) = \chi_0(123) + \int G(14)G(52)\frac{\delta\Sigma(45)}{\delta G(67)}\chi(673)d(4567). \tag{3.48}$$

In order to make effective use of this equation we need an approximation for $\Sigma$. It is perhaps worth stressing again that in the case of strongly singular potentials we cannot employ the Hartree–Fock approximation for the self–energy. Some modified approximation such as that provided by the correlated iteration procedure is essential.

From the *correlated Hartree approximation*

$$\Sigma_{CH}(12) := \delta(12) \int \bar{\Phi}(13)\rho(3)d(3), \qquad (3.49)$$

since

$$\frac{\delta \Sigma_{CH}}{\delta G(34)} = \pm i\delta(12)\bar{\Phi}(23)\delta(34),$$

noting (3.48) we conclude that

$$\chi(123) = \chi_0(123) + \int \chi_0(124)\bar{\Phi}(45)\chi(553)d(45).$$

By specialization this shows that the equation for (3.43) takes the form

$$\chi(12) = \chi_0(12) + \int \chi_0(13)\bar{\Phi}(34)\chi(42)d(34), \qquad (3.50)$$

where

$$\chi_0(12) := \pm iG(12)G(21) = \Pi_0(12).$$

We shall say that the solution of (3.50) is the response function in *correlated random–phase approximation*. In the standard random–phase approximation, the smoothed potential, $\bar{\Phi}$, would be replaced by the bare potential, $\Phi$. If we employ the correlated Hartree–Fock approximation (3.12) for $\Sigma$ in (3.48) we are led to the equation

$$\chi(123) = \chi_0(123)\pm \qquad (3.51)$$

$$\pm i \int G(14)\bar{\Phi}(45)[G(42)\chi(553) \pm G(52)\chi(453)]d(45).$$

We note in Chapter 2 that the poles of the two–particle propagators define the spectrum of collective excitations. The relation (3.46) shows that these propagators and the density response functions have the same poles. Since the equations for the response functions are usually simpler than those for the 2–propagators, the response functions are commonly considered when the collective excitations spectrum is of interest. We describe two possible approaches to finding this spectrum.

1. Substitute an approximation for the self–energy, $\Sigma$, in (3.48) which results in an equation for the *generalized* response function. If possible, specialize to obtain an equation for the density response function. Except for usual cases, this specialization is feasible only when the Hartree approximation is employed for the self–energy, but then as follows from (3.51) this specialization will be impossible in the next approximation. Further, to proceed from the first approximation one needs to solve the resulting equation for the density response function. Because of the product in the integrand of the right–hand side of (3.50) this can be quite difficult unless one uses merely a random–phase type of approximation. In this latter case we find a solution for $\chi(12)$ in the form

$$\chi_{CRP} = \frac{\chi_0}{1 - \chi_0\bar{\Phi}}, \qquad (3.52)$$

where the subscript, $CRP$, means *correlated random–phase* approximation. The spectrum of collective excitations is then approximated by the zeros of the equation

$$1 - \chi_0 \bar{\bar{\Phi}} = 0. \tag{3.53}$$

However, it is not always easy to solve (3.51) for the density response function.

**2.** A second approach to finding the spectrum of the collective excitations begins by seeking an approximate iterative solution of (3.48) in the form of a series in powers of $\bar{\bar{\Phi}}$. Since the poles of $\chi$ are the zeros of $\chi^{-1}$, the spectrum is characterized by the equation $\chi^{-1} = 0$ after $\chi$ is replaced by a polynomial in $\bar{\bar{\Phi}}$ of degree equal to the order of approximation one is using for $\chi$. For example, taking

$$\chi \simeq \chi_0 + \chi_0 \bar{\bar{\Phi}} \chi_0 \tag{3.54}$$

leads to

$$\chi^{-1} \simeq \left(1 - \chi_0 \bar{\bar{\Phi}}\right) \chi_0^{-1}, \tag{3.55}$$

when the infinite series for $\chi^{-1}$ is truncated at the first order in $\bar{\bar{\Phi}}$. The zeros of this equation are the same as those of equation (3.53). Notice that it might happen that (3.54) has no poles even though (3.55) has zeros! This could result because the truncation meant that these two equations are not strictly equivalent.

In order to carry this second approach through effectively we need to have at our disposal a "regular" perturbation procedure, – that is a procedure which does not terminate abruptly due to the occurrence of singularities. We can define such a procedure by employing the same trick for inverting operators as we used in our correlated iteration theory for the doubling function and self–energy in Section 3.2.

We introduce an integral operator, $\mathbf{Z}$, on $H^3$ such that $\mathbf{Z}f$ is defined by

$$\mathbf{Z}f(123) := \int G(14)G(52)\frac{\delta\Sigma(45)}{\delta G(67)}f(673)d(4567). \tag{3.56}$$

Then the basic equation, (3.48), for the generalized response function can be written as

$$(\mathbf{I} - \mathbf{Z})\chi(123) = \chi_0(123). \tag{3.57}$$

This is equivalent to

$$\chi(123) = (\mathbf{I} - \mathbf{Z})^{-1}\chi_0(123), \tag{3.58}$$

in which the operator can be expanded in the series,

$$(\mathbf{I} - \mathbf{Z})^{-1} = \sum_{k=0}^{\infty} \mathbf{Z}^k. \tag{3.59}$$

It is clear from (3.56) that $\mathbf{Z}$ is a functional of the self–energy. We symbolize this by

$$\mathbf{Z} = \mathbf{Z}[\Sigma], \tag{3.60}$$

and denote the operator $k$–approximation to $\mathbf{Z}$ by

$$\mathbf{Z}_k := \mathbf{Z}[\Sigma]_k, \qquad \mathbf{Z}_0 := \mathbf{I}, \tag{3.61}$$

where $\Sigma_k$ denotes the $k$–approximation for the self–energy in the correlated iterative scheme. The $k$–approximation for the generalized response function is then $\chi_k$, defined by

$$\chi_k(123) := \sum_{p=0}^{k} \mathbf{Z}_k^p \chi_0(123). \tag{3.62}$$

In the previous Section we showed how to obtain successive approximations to the self–energy by means of correlated iterations, so following the scheme

$$\Sigma_k \to \mathbf{Z}_k \to \chi_k \tag{3.63}$$

we may obtain successive approximations to the generalized response function. According to (3.43) the $k$–approximation to the density response function is obtained by the specialization

$$\chi_k(12) = \chi_k(112). \tag{3.64}$$

For example, with $k = 1$ in (3.62) we have

$$\chi_1(123) = \chi_0(123)\pm$$

$$\pm i \int G(14)\bar{\Phi}(45)[G(42)\chi_0(553) \pm G(52)\chi_0(453)]d(45).$$

It follows that

$$\chi_1(12) = \chi_0(12) + \int \chi_0(13)\bar{\Phi}(34)\chi_0(42)d(34)\pm \tag{3.65}$$

$$\pm \int \chi_0(123)\bar{\Phi}(34)\chi_0(214)d(34).$$

The schemes (3.25) and (3.63) provide a recursive procedure for obtaining a sequence of successive approximations to the density response function in powers of $\bar{\Phi}$ which define $\chi$ asymptotically. The poles of the function so–defined, or the zeros of $\chi^{-1}$, characterize the spectrum of collective excitations of the system.

## 3.5   Correlation Function

From the above analysis it follows that the smoothing function and the pair correlation function are intimately related with each other. Really, comparing the definitions (3.5) and (3.38), we see that the smoothing function can be identified with an approximation for the pair correlation function. Therefore, to make the correlated iteration theory complete, we need to present a way for finding approximate correlation functions.

The latter, according        fined as

$$\qquad (t_{12} > 0). \tag{3.66}$$

If we substitute in (3.66)

$$G_2^0(1221) = -\rho(1)\rho(2) \pm G(12)G(21) \,,$$

then we get

$$g_0(12) = 1 \mp \frac{G(12)G(21)}{\rho(1)\rho(2)} \,. \tag{3.67}$$

A general equation for $g(12)$ can be obtained employing in (3.13) definition (3.66), which gives

$$g(12) = g_0(12) + \int \frac{G(13)G(41)}{\rho(1)\rho(2)} \cdot \frac{\delta\Sigma(34)}{\delta G(56)} \times$$

$$\times \left[ G(56)G(22) - g(52) \int D(527)G(76)d(7) \right] d(3456), \tag{3.68}$$

keeping in mind the condition $t_{12} > 0$. Using in (3.68) approximate expressions for $\Sigma$ and $D$, we then come to a set of approximate equations defining $g(12)$. For instance, putting $\Sigma = \Sigma_0 = 0$ in (3.68) immediately returns us to (3.67). If we take for $\Sigma$ and $D$ the first approximations in (3.12) and (3.37), respectively, then (3.68) transforms to

$$g(12) = g_0(12) - \int A(1234)g(34)\Phi(34)d(34) \,, \tag{3.69}$$

where we keep only the terms up to first order in $\Phi$ and

$$A(1234) := i\frac{G(13)G(41)}{\rho(1)\rho(2)} \left\{ g(32)\left[2 - g(32)\right]G_2^0(3224) - G(34)G(22) \right\} \pm$$

$$\pm i\frac{G(13)G(31)}{\rho(1)\rho(2)} \left\{ g(42)\left[2 - g(42)\right]G_2^0(4224) - G(44)G(22) \right\} \,.$$

Equation (3.69) defines first–order approximation for $g(12)$.

Let us notice that (3.69) cannot be iterated starting with $g_0(12)$ in the integrand, since then integral would diverge because of singularity in $\Phi(34)$. A simplification is possible only for $A(1234)$. If we put in the latter $g = 1$, then

$$A(1234) = i\frac{G(13)G(24)}{\rho(1)\rho(2)}G_2^0(3421) \,.$$

Taking $g = 1$ implies that the Hartree approximation is used for which

$$G_2^0(1234) \cong G(14)G(23), \qquad g_0(12) \cong 1 \,.$$

With this in mind, we can make a further simplification getting

$$A(1234) \cong i\frac{G(13)G(31)G(24)G(42)}{\rho(1)\rho(2)} \,.$$

If we substitute in (3.68) the second–order approximations for $\Sigma_2$ and $D_2$, then we come to an equation defining the second–order approximation for the correlation function $g(12)$. However, this equation is rather cumbersome. Instead, we may improve the first–order approximation for $g(12)$ in the following way. We can consider (3.69) as an asymptotic expansion in powers of $\Phi$. An effective limit of an asymptotic series can be defined by using the self–similar approximation theory [85–97]. Employing the algebraic variant of this theory [98-100], we obtain

$$g(12) = g_0(12)\exp\left\{-\int A(1234)\frac{g(34)}{g_0(12)}\Phi(34)d(34)\right\} . \qquad (3.70)$$

This equation, as is explained above, because of divergences, cannot be iterated starting with $g_0(12)$. One direct option would be to solve (3.70) numerically for a given interaction potential $\Phi$. Another way can be by iterating the right–hand side of (3.70) starting with an approximate $g_{app}(12)$ taking into account interparticle correlations. For example, we may start [82] with

$$g_{app}(12) = |\Psi(12)|^2 ,$$

where $\Psi(12)$ is a pair wave function normalized so that to satisfy the condition

$$g_{app}(12) \to 1, \qquad |x_1 - x_2| \to \infty ,$$

similar to (3.40). In this way, the correlated iteration theory is completed.

# Appendix

It was mentioned in Preface that statistical mechanics is the most general physical theory. Here we concretize this statement.

From the mathematical point of view, one theory is more general than another if the latter can be derived from the former by means of some mathematical operations, but not conversely. A particular theory can be considered as a limiting case of a more general theory. The corresponding limit can be defined explicitly in an unambiguous mathematical sense. The procedure of taking a limit is, of course, to be accompanied by the related physical interpretation.

The classification of physical theories can be done by considering three parameters: velocity of light, $c$, Plank constant, $\hbar$, and Boltzmann constant, $k_B$. The choice of these parameters is caused by the fact that each of them is related to a principal physical notion. The light velocity, according to the Einstein postulate, is the same in all reference systems. The Plank constant defines a minimal quant of action. And the Boltzmann constant connects the entropy with a statistical distribution.

A theory with the finite velocity of light, $c$, is called *relativistic*. The limit $c \to \infty$ leads to a *nonrelativistic* theory. A theory containing the Plank constant, $\hbar$, is termed *quantum*. Taking the limit $\hbar \to 0$, one passes to a *classical* theory. A theory dealing with probability distributions and, thus, containing the Boltzmann constant, $k_B$, is named *statistics*. The limit $k_B \to 0$ reduces the latter to a particular kind of theory for which there is no a commonly accepted name. For such theories, we use below the word "mechanics", although this term is more general and is applicable not only to theories with $k_B = 0$. So that the general name "mechanics" is to be considered not separately but in the combination with other terms characterizing a theory. The whole theoretical physics can be subdivided into 8 theories according as the principal parameters from the triplet $\{c, \hbar, k_B\}$ are finite or some limits are taken. These theories are as follows:

(1) Relativistic Quantum Statistics: $\{c, \hbar, k_B\}$;

(2) Nonrelativistic Quantum Statistics: $\{c \to \infty, \hbar, k_B\}$;

(3) Relativistic Classical Statistics: $\{c, \hbar \to 0, k_B\}$;

(4) Nonrelativistic Classical Statistics: $\{c \to \infty, \hbar \to 0, k_B\}$;

(5) Relativistic Quantum Mechanics: $\{c, \hbar, k_B \to 0\}$;

(6) Nonrelativistic Quantum Mechanics: $\{c \to \infty, \hbar, k_B \to 0\}$;

(7) Relativistic Classical Mechanics: $\{c, \hbar \to 0, k_B \to 0\}$;

(8) Nonrelativistic Classical Mechanics: $\{c \to \infty, \hbar \to 0, k_B \to 0\}$.

As is clear from this classification, Relativistic Quantum Statistics is the most general theory from which all other theories can be derived by means of appropriate limits. If not to subdivide the theories onto relativistic and nonrelativistic, then Quantum Statistics is the most general physical theory incapsulating all other theories.

# References


[1] J. Von Neumann, *Matematische Grundlagen der Quantenmechanik* (Springer, Berlin, 1932).

[2] J. Von Neumann, *Mathematical Foundation of Quantum Mechanics* (Princeton Univ., Princeton, 1955).

[3] P. A. M. Dirac, *Principles of Quantum Mechanics* (Oxford Univ., Oxford, 1935).

[4] P. O. Löwdin, Phys. Rev. **97**, 1474 (1955).

[5] R. McWeeny, Rev. Mod. Phys. **32**, 335 (1960).

[6] D. Ter Haar, Rep. Prog. Phys. **24**, 304 (1961).

[7] C. N. Yang, Rev. Mod. Phys. **34**, 694 (1962).

[8] A. J. Coleman, Rev. Mod. Phys. **35**, 668 (1963).

[9] A. J. Coleman, J. Math. Phys. **6**, 1425 (1965).

[10] F. Sasaki, Phys. Rev. B **138**, 1338 (1965).

[11] R. P. Feynman, *Statistical Mechanics* (Benjamin, Reading, 1972).

[12] E. R. Davidson, *Reduced Density Matrices in Quantum Chemistry* (Academic, New York, 1976).

[13] K. Blum, *Density Matrix Theory and Applications* (Plenum, New York, 1981).

[14] R. M. Erdahl and V. H. Smith, eds., *Density Matirces and Density Functionals* (Reidel, Dordrecht, 1987).

[15] D. Ter Haar, *Elements of Statistical Mechanics* (Rinehart, New York, 1954).

[16] L. D. Landau and E. M. Lifshitz, *Statistical Physics* (Pergamon, Oxford, 1958).

[17] J. E. Mayer and M. G. Mayer, *Statistical Mechanics* (Wiley, New York, 1977).

[18] O. Penrose, Rep. Prog. Phys. **42**, 1937 (1979).

[19] V. I. Yukalov, Phys. Rep. **208**, 395 (1991).

[20] A. Einstein, B. Podolsky, and N. Rosen, Phys. Rev. **47**, 777 (1935).

[21] B. d'Espagnat, *Conceptual Foundations of Quantum Mechanics* (Benjamin, Reading, 1976).

[22] B. d'Espagnat, Phys. Rep. **110**, 201 (1984).

[23] V. I. Yukalov, Mosc. Univ. Phys. Bull. **25**, 49 (1970).

[24] V. I. Yukalov, Mosc. Univ. Phys. Bull. **26**, 22 (1971).

[25] F. A. Berezin, *Method of Second Quantization* (Academic, Ney York, 1966).

[26] G. G. Emch, *Algebraic Methods in Statistical Mechanics and Quantum Field Theory* (Wiley, New York, 1972).

[27] O. Bratteli and D. Robinson, *Operators Algebras and Quantum Statistical Mechanics* (Springer, New York, 1979).

[28] A. J. Coleman and V. I. Yukalov, Mod. Phys. Lett. B **5**, 1679 (1991).

[29] A. J. Coleman and V. I. Yukalov, Nuovo Cimento B **107**, 535 (1992).

[30] A. J. Coleman and V. I. Yukalov, Nuovo Cimento B **108**, 1377 (1993).

[31] A. J. Coleman and V. I. Yukalov, Int. J. Mod. Phys. B **10**, 3505 (1996).

[32] F. Schlögl, Phys. Rep. **62**, 267 (1980).

[33] J. W. Gibbs, *Collected Works*, Vol. 1 (Longmans, New York, 1928).

[34] J. W. Gibbs, *Collected Works*, Vol. 2 (Longmans, New York, 1931).

[35] D. Ruelle, *Statistical Mechanics* (Benjamin, New York, 1969).

[36] H. Mori, J. Phys. Soc. Japan **11**, 1029 (1956).

[37] J. A. McLennan, Phys. Rev. **115**, 1405 (1959).

[38] N. N. Bogolubov, *Problems of Dynamical Theory in Statistical Physics* (North–Holland, Amsterdam, 1962).

[39] N. N. Bogolubov, *Lectures on Quantum Statistics*, Vol. 2 (Gordon and Breach, New York, 1970).

[40] D. N. Zubarev, *Nonequilibrium Statistical Thermodynamics* (Consultants Bureau, New York, 1974).

[41] D. Ter Haar, *Lectures on Selected Topics in Statistical Mechanics* (Pergamon, Oxford, 1977).

[42] R. Kubo, M. Toda, and N. Hashitsume, *Nonequilibrium Statistical Mechanics* (Springer, Berlin, 1991).

[43] M. D. Girardeau, Phys. Rev. B **139**, 500 (1965).

[44] M. D. Girardeau, J. Math. Phys. **10**, 1302 (1969).

[45] V. I. Yukalov, Mosc. Univ. Phys. Bull. **27**, 59 (1972).

[46] V. I. Yukalov, Proc. Mosc. Univ. Phys. **59**, 91 (1972).

[47] M. Ozaki, J. Math. Phys. **26**, 1514 (1985).

[48] R. J. Glauber, Phys. Rev. **130**, 2529 (1963).

[49] J. R. Klauder and E. C. Sudarshan, *Fundamentals of Quantum Optics* (Benjamin, New York, 1968).

[50] J. R. Klauder and B. S. Skagerstam, *Coherent States* (World Scientific, Singapore, 1985).

[51] A. Inovata, H. Kuratsuji, and C. Gerry, *Path Integrals and Coherent States of SU(2) and SU(1,1)* (World Scientific, Singapore, 1992).

[52] V. L. Bonch–Bruevich and S. V. Tyablikov, *Green Functions Method in Statistical Mechanics* (North–Holland, Amsterdam, 1962).

[53] L. P. Kadanoff and G. Baym, *Quantum Statistical Mechanics* (Benjamin, New York, 1962).

[54] A. A. Abrikosov, L. P. Gorkov, and I. E. Dzyaloshinskii *Methods of Quantum Field Theory in Statistical Physics* (Pentice–Hall, New Jersey, 1963).

[55] D. A. Kirzhnits, *Field–Theoretical Methods in Many–Body Systems* (Pergamon, Oxford, 1967).

[56] S. V. Tyablikov, *Methods in Quantum Theory of Magnetism* (Plenum, New York, 1967).

[57] E. M. Lifshitz and L. P. Pitaevskii, *Statistical Physics* (Pergamon, Oxford, 1980).

[58] J. Linderberg and Y. Öhrn, *Propagators in Quantum Chemistry* (Academic, London, 1973).

[59] V. I. Yukalov, Ann. Physik **38**, 419 (1981).

[60] A. Dalgarno and J. Lewis, Proc. Roy. Soc. London A **233**, 70 (1955).

[61] A. Dalgarno and J. Lewis, Proc. Roy. Soc. London A **69**, 628 (1956).

[62] J. C. Slater and J. G. Kirkwood, Phys. Rev. **37**, 682 (1931).

[63] A. Dalgarno and A. Stewart, Proc. Roy. Soc. London A **238**, 269 (1956).

[64] L. Hedin and J. Lundquist, Solid State Phys. **23**, 1 (1969).

[65] P. Streitenberger, Phys. Lett. A **106**, 57 (1984).

[66] P. Streitenberger, Phys. Status Solidi B **125**, 681 (1984).

[67] T. Hidaka, Phys. Rev. B **48**, 9313 (1993).

[68] R. Del Sole, L. Reining, and R. Godby, Phys. Rev. B **49**, 8024 (1994).

[69] P. Bobbert and W. Van Haeringen, Phys. Rev. B **49**, 10326 (1994).

[70] R. D. Mattuck, *Guide to Feynman Diagrams in the Many–Body Problem* (McGraw–Hill, London, 1967).

[71] V. I. Yukalov, Theor. Math. Phys. **26**, 274 (1976).

[72] V. I. Yukalov, Theor. Math. Phys. **28**, 652 (1976).

[73] V. I. Yukalov, Physica A **89**, 363 (1977).

[74] V. I. Yukalov, in *Selected Topics in Statistical Mechanics*, edited by N. N. Bogolubov (Joint Inst. Nucl. Res., Dubna, 1978), p.437.

[75] V. I. Yukalov, Phys. Lett. A **81**, 249 (1981).

[76] V. I. Yukalov, Phys. Lett. A **81**, 433 (1981).

[77] V. I. Yukalov, Phys. Lett. A **85**, 68 (1981).

[78] V. I. Yukalov, Physica A **108**, 402 (1981).

[79] V. I. Yukalov, Phys. Lett. A **125**, 95 (1987).

[80] V. I. Yukalov, Physica A **141**, 352 (1987).

[81] V. I. Yukalov, in *Problems in Quantum Field Theory*, edited by N. N. Bogolubov (Joint Inst. Nucl. Res., Dubna, 1987), p.62.

[82] V. I. Yukalov, Int. J. Theor. Phys. **28**, 1237 (1989).

[83] V. I. Yukalov, Nuovo Cimento A **103**, 1577 (1990).

[84] V. I. Yukalov, Phys. Rev. A **42**, 3324 (1990).

[85] V. I. Yukalov, *Renormalization Group in Statistical Physics: Field–Theory and Iteration Formulations*, JINR Commun. P17–88–893, Dubna (1988).

[86] V. I. Yukalov, Int. J. Mod. Phys. B **3**, 1691 (1989).

[87] V. I. Yukalov, Physica A **167**, 833 (1990).

[88] V. I. Yukalov, Proc. Lebedev Phys. Inst. **188**, 297 (1991).

[89] V. I. Yukalov, J. Math. Phys. **32**, 1235 (1991).

[90] V. I. Yukalov, J. Math. Phys. **33**, 3994 (1992).

[91] E. P. Yukalova and V. I. Yukalov, Bulg. J. Phys. **19**, 12 (1992).

[92] E. P. Yukalova and V. I. Yukalov, Phys. Lett. A **175**, 27 (1993).

[93] E. P. Yukalova and V. I. Yukalov, J. Phys. A **26**, 2011 (1993).

[94] E. P. Yukalova and V. I. Yukalov, Phys. Scr. **47**, 610 (1993).

[95] V. I. Yukalov, Int. J. Mod. Phys. B **7**, 1711 (1993).

[96] V. I. Yukalov and E. P. Yukalova, Int. J. Mod. Phys. B **7**, 2367 (1993).

[97] V. I. Yukalov and E. P. Yukalova, Physica A **198**, 573 (1993).

[98] V.I. Yukalov and S. Gluzman, Phys. Rev. Lett. **79**, 333 (1997).

[99] S. Gluzman and V.I. Yukalov, Phys. Rev. E **55**, 3983 (1997).

[100] V.I. Yukalov and S. Gluzman, Phys. Rev. E **55**, 6552 (1997).